THE HISTORY TREE

*Moments in the Lifetime
of a Memorable Tree*

Edited by
Janet Cochrane

NORTH YORKSHIRE MOORS ASSOCIATION

First published in Great Britain in 2018

British Library Cataloguing-in-Publication Data
A CIP record for this title is available from the British Library

ISBN 978 0 9565779 5 5

Front cover: The History Tree in front of the Moors Centre c. 2005.
Back cover: The History Tree Stump

NORTH YORKSHIRE MOORS ASSOCIATION
4 Station Road
Castleton
Whitby
North Yorkshire YO21 2EG

NYMA is a Charitable Incorporated Organisation, registration no. 1169240
This book was part-funded by the Heritage Lottery Fund and the Land of Iron project

Digital Artwork by Ian Dashper
Printed and Bound by W&G Baird, Greystone Press, Caulside Drive, Co. Antrim, Northern Ireland

heritage
lottery fund
LOTTERY FUNDED

LAND OF
IRON
1837 1929

David Ross Foundation
Broadening Horizons

CONTENTS

Moments in Time

Engraving of Danby Lodge in the mid-1800s with the young History Tree to the right

The North York Moors National Park

Places mentioned in the book

Middlesbrough
Kirkleatham
Hummersea
Street House
Boulby
Staithes
North Ormesby
Teesmouth
Eston
GUISBOROUGH
North Skelton
LOFTUS
Runswick Bay
Marton
Newton Mulgrave
Lythe
Sandsend
Highcliffe Nab
Mulgrave
WHITBY
Roseberry Topping
Danby Beacon
Great Ayton
Easby Moor
Stonegate
Commondale
Bannial Flatt Farm
Danby
Lealholm
STOKESLEY
Park Nab
Castleton
Eskdale
Sleights
Ingleby Greenhow
Danby Dale
Grosmont
Ingleby Incline
Fryupdale
Robin Hood's Bay
Botton
Boggle Hole
Swainby
Wainstones
Danby High Moor
Beck Hole
Ravenscar
Scugdale
Urra Moor
Ralph Cross
Millennium Stone
Goathland
Blakey Ridge
Lilla Cross
Fylingdales Moor
Rosedale
Bilsdale
Hole of Horcum
Hackness
Hutton-le-Hole
Levisham
Cropton
Dalby
Rievaulx Abbey
KIRKBYMOORSIDE
SCARBOROUGH
Sutton Bank
Kirkdale
Lake Gormire
Helmsley
Wykeham
White Horse
PICKERING
Brompton-by-Sawdon
THIRSK
Ampleforth
Kilburn
Coxwold
Newburgh Priory
MALTON

DEDICATION

*The timeless nature of the North York Moors National Park - its
landscape, natural and cultural heritage - has given pleasure to
countless millions over the years, and will continue to do so.
This book is dedicated to all those people of the past, present and future
whose efforts protect the Park's irreplaceable qualities.*

FOREWORD

by Viscountess Downe

My first memory of Danby Lodge reaches back to September 1966, just a year after my marriage to my late husband John, when we came up for the annual grouse-shooting season.

We soon realised that these historic buildings, set in the beautiful surroundings of the Esk Valley, would make an ideal information centre for the National Park. In May 1976 we were delighted when this vision was finally realised with the Lodge opening its doors to visitors for the first time. Since then the Moors Centre has provided a focal point for the tens of thousands of visitors who come each year from near and far to experience and enjoy the heather moors and learn more about the history and culture of this fascinating area.

Sadly, the magnificent copper beech tree that graced the front of the Lodge has now been lost. However, I was heartened to see the memory of this noble tree living on through the imaginative History Tree Project. In 2010 the North Yorkshire Moors Association placed an inscribed commemorative plate where the tree once stood. And now the publication of this attractive book – a varied collection of essays and photographs – brings new life and meaning to the story of the Tree and will inspire a greater understanding and appreciation of our irreplaceable National Park and the natural world which we all share.

Diana Downe

Wykeham Abbey
North Yorkshire

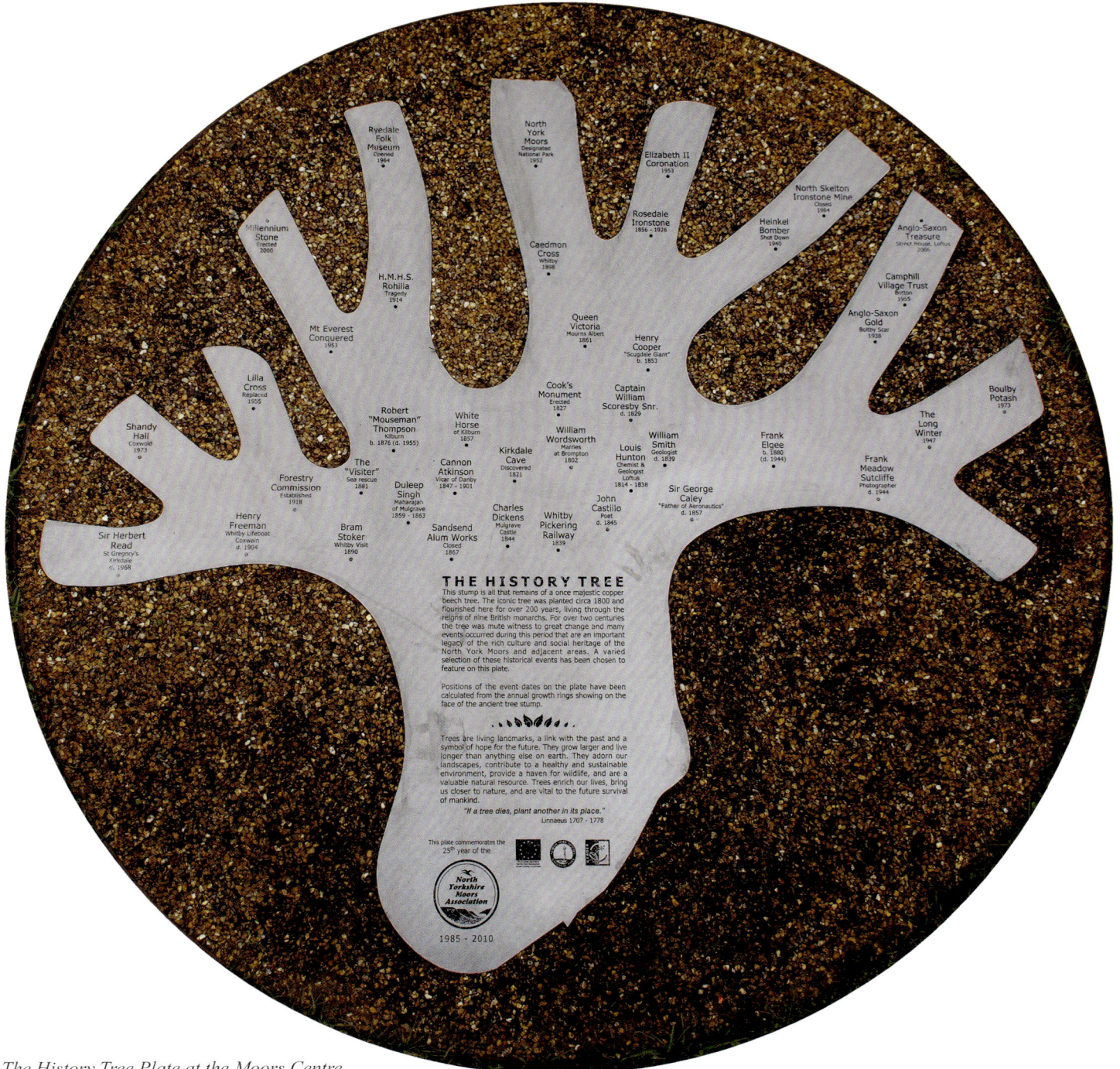

Ryedale Folk Museum
Opened
1964

North York Moors
Designated National Park
1952

Elizabeth II Coronation
1953

North Skelton Ironstone Mine
Closed
1964

Millennium Stone
Erected
2000

Caedmon Cross
Whitby
1898

Rosedale Ironstone
1856 - 1926

Heinkel Bomber
Shot Down
1940

Anglo-Saxon Treasure
Street House, Loftus
2006

H.M.H.S. Rohilla
Tragedy
1914

Camphill Village Trust
Botton
1955

Mt Everest Conquered
1953

Queen Victoria
Mourns Albert
1861

Henry Cooper
"Scugdale Giant"
b. 1853

Anglo-Saxon Gold
Boltby Scar
1938

Lilla Cross
Replaced
1955

Cook's Monument
Erected
1827

Captain William Scoresby Snr.
d. 1829

Boulby Potash
1973

Shandy Hall
Coxwold
1973

Robert "Mouseman" Thompson
Kilburn
b. 1876 (d. 1955)

White Horse
of Kilburn
1857

William Wordsworth
Marries at Brompton
1802

William Smith
Geologist
d. 1839

Frank Elgee
b. 1880
(d. 1944)

The Long Winter
1947

Forestry Commission
Established
1918

The "Visiter"
Sea rescue
1881

Kirkdale Cave
Discovered
1821

Louis Hunton
Chemist & Geologist
Loftus
1814 - 1838

Frank Meadow Sutcliffe
Photographer
d. 1944

Henry Freeman
Whitby Lifeboat
Coxswain
d. 1904

Cannon Atkinson
Vicar of Danby
1847 - 1901

Duleep Singh
Maharajah of Mulgrave
1859 - 1863

Charles Dickens
Mulgrave Castle
1844

John Castillo
Poet
d. 1845

Sir George Caley
"Father of Aeronautics"
d. 1857

Sir Herbert Read
St Gregory's
Kirkdale
d. 1968

Bram Stoker
Whitby Visit
1890

Sandsend Alum Works
Closed
1867

Whitby Pickering Railway
1839

THE HISTORY TREE

This stump is all that remains of a once majestic copper beech tree. The iconic tree was planted circa 1800 and flourished here for over 200 years, living through the reigns of nine British monarchs. For over two centuries the tree was mute witness to great change and many events occurred during this period that are an important legacy of the rich culture and social heritage of the North York Moors and adjacent areas. A varied selection of these historical events has been chosen to feature on this plate.

Positions of the event dates on the plate have been calculated from the annual growth rings showing on the face of the ancient tree stump.

Trees are living landmarks, a link with the past and a symbol of hope for the future. They grow larger and live longer than anything else on earth. They adorn our landscapes, contribute to a healthy and sustainable environment, provide a haven for wildlife, and are a valuable natural resource. Trees enrich our lives, bring us closer to nature, and are vital to the future survival of mankind.

"If a tree dies, plant another in its place."
Linnaeus 1707 - 1778

This plate commemorates the 25th year of the

North Yorkshire Moors Association

1985 - 2010

The History Tree Plate at the Moors Centre

Introduction

One winter's day in early 2007, a truly awful sound echoed around the Esk Valley. From the front of Danby Lodge - The Moors Centre - across the River Esk, over Duck Bridge on to Danby Castle and beyond, incessantly it rasped through the morning air: chainsaws! And whenever this dreadful sound is heard, you can guarantee there is a tree on the receiving end of it.

So it was, in just a few hours, two hundred years of life, formed from the air, rain, soil and sunshine through the passage of time, a magnificent copper beech tree was no more, reduced to a tangle of severed branches, twigs and slices of trunk spread across the ground. Fears over safety had sounded the death knell, demanding it be cut down, and with it went a delightful and irreplaceable feature in the landscape and the setting of the Lodge. It was no-one's fault, no-one to blame, just that – sadly - nothing is forever. But the memory lives on.

Such was the sense of loss of this beautiful tree, that to celebrate the 25th anniversary of the founding of the North Yorkshire Moors Association (NYMA), a steel plate etched with events during the decades of its life was placed where the tree once stood. At first it was fixed to the tree stump. Later, when the wood decayed, the remains of the tree were removed and the 'History Tree' plate was set permanently into the ground.

Looking at the empty space today, where the vast branches once stretched across a ha-ha, offering cool shade on summer days, triggers many fond memories for me.

The Moors Centre, Danby Lodge 2018

I first encountered the History Tree 40 years ago, when I came to Danby Lodge to be interviewed for the role of Assistant Director of the National Park Centre. No-one could fail to have been impressed by its imposing presence in front of the buildings. As with any ancient tree it challenged the imagination to contemplate the events it witnessed, or that occurred in the wider world, over its years.

Our tree would have shared this corner of the moors with all manner of events and changes: look-outs passing up or down to Danby Beacon at the time of the Napoleonic wars; the first trains, first motor cars, first aeroplanes and first electric lights, which would begin their insidious assault on the darkness, to mention just a few.

Seasons would come and go, the wind would sound the same, the rain, snow and frost would be the same; the sparkling of moonlight and the stars, the same. The sounds of the birds - the owls, the cuckoo and the curlew, the same; the sounds of war planes; it heard them too, not just during the Second World War, but through the Cold War, as thunderous jets on manoeuvres before the fall of the Soviet Union shook the very foundations of nearby buildings with the shock-waves of their deafening roar.

There are indeed many points in the tree's story that could have been chosen to commemorate on the Plate. Practically, someone had to make that choice, and no doubt each of us would have had a different view.

I will suggest four I might have argued to be added. The first two are rather technical, but hugely important for the National Park, wildlife and the countryside; the third just a lovely experience and the last an event that includes a bit of self-congratulation.

My first choice: the Government scrapping the grant payments for converting heather moorland into grassland in the early 1980s, which had been a devastatingly bad idea for the landscape and all that lived on the moorlands concerned. Having been a part of the campaign aimed at stopping this seriously harmful practice, it was, for me, a pivotal moment in the drive to protect the moorland.

Moorland conversion to grassland

Secondly: the adoption of the EU Habitats Directive (1992) and its associated Regulations, for the key part they have played in establishing Special Areas of Conservation and Special Protection Areas for Birds which embrace much of the moorland. These rules placed the onus on anyone proposing change to show that it wouldn't cause harm to the environment, rather than the hitherto other way round.

Thirdly, reflecting the very best of the National Park Movement: the Festival of National Parks at Chatsworth in 1989, held to celebrate the 40th Anniversary of the legislation which founded the National Parks. A unique occasion, sadly unlikely ever to be repeated, saw the National Parks putting on a large country fair celebration on a beautiful September day at Chatsworth House in Derbyshire, honoured by the presence of Diana, Princess of Wales.

In a complex network of journeys, representatives of each Park visited over 90 cities throughout England as they worked their way across the country, converging on the centre of Birmingham for a massed rally through the streets, before travelling on to Chatsworth House for the Festival. Staff from our own Moors left Rievaulx Abbey dressed as monks and a nun, towing a two-dimensional sheep on wheels and carrying the banner for the North York Moors.

It was a spectacular demonstration of national park unity and collective sense of purpose.

Finally, I would also have added the foundation of NYMA as a tree-ring position on the plate.

Most National Parks in the United Kingdom have a respective 'friends of the area'. Run by enthusiastic and committed individuals, mostly volunteers, they provide mechanisms to both support the Park Authorities and at times to challenge and hold them to account when the principles which underpin the well-being of the parks appear at risk of being compromised.

What exactly does the North Yorkshire Moors Association actually do, and what has it achieved over the

The Festival of National Parks, Chatsworth House

years? The answer to this question gives me the golden opportunity to unashamedly applaud the Association and to emphasise the importance of membership support for its work. However, a few words and one or two examples can never come anywhere near reflecting its full value.

Since the History Tree was felled, many of the Association's important 'moments in time' post-date the tree's demise. This in no way undermines the logic of mentioning them here. Anyway, the roots of the tree are still down there, decaying back into the earth from which they came, as is the natural order of things.

The 'voice' of NYMA was first heard in 1985, when the Rt. Hon. William Waldegrave, Minister of the Environment, spoke at the inaugural Annual General Meeting of the Association in Hutton-le-Hole.

The vision of the Association's founders, Dr Don Tilley, Major Peter Walker, Derek Statham and Gerald McGuire, had been firmly set to engage, inform, support, enjoy and sometimes challenge in the pursuit of safeguarding the future of the North York Moors landscape and culture.

In the face of constant pressures on the countryside their mission could not have been more timely. For a while, we saw the highpoints of enthusiasm and momentum for positive action for the environment in government and local authorities. This didn't last. In the face of economic pressures, the roles of leading bodies such as the Countryside Commission, champions of the original Countryside Stewardship scheme, Country Parks, Heritage Coasts and Community Forests amongst a suite of other initiatives - and English Nature, guardian of the natural world for the nation - were forced into decline. With narrower remits and cuts in resources they merged and unacceptably withered into virtually nothing as the practically toothless Natural England, an organisation staffed by dedicated people at local level but with little central support to do the job that really needs to be done.

Only with concerted pressure from the public, through the continued effort of well-informed and motivated individuals and organisations, can we help to moderate the risk of constant attrition and irreversible loss of our precious countryside.

But the Moors aren't just a battleground of gloom. Cliché it might be, but 'laugh and the world laughs with you, cry and you cry alone', is so very true. Several books have been published by NYMA, the quarterly 'Voice of the Moors' magazine keeps members informed; there are walks, events, talks and visits. Alongside these, direct conservation

Lapwing – the Moors are a stronghold for wildlife

projects have seen tree-planting in Park Wood and nurturing of juniper seedlings to restore this lovely tree to the Park, for example. Most spectacularly, and of great national significance, a leading role in the collaborative Cornfield Flowers Project with CCT (a conservation charity), the National Park Authority and Ryedale Folk Museum has seen many rare plants of arable fields pulled back from the brink. Without the help of the Association the prospect for some would have been even more bleak.

Looking after the countryside is a bitter-sweet experience. Support for the Moorsbus, a delightful car-free and care-free transport to and around the Moors, is a joy, while at the other end of the spectrum, influencing local and national policies and planning matters, on which the character of the Park ultimately depends, can be a mind-numbing and time-consuming task.

When it comes to inappropriate major developments, the going gets really tough. Along with the Campaign for National Parks, the Association successfully encouraged the Secretary of State to overturn the National Park Authority's decision to allow an extension to Spaunton Quarry, a potential precedent in breach of national policy to avoid the operation of such works in the Parks.

Recent times have seen the National Park Authority, under considerable pressure, narrowly approve by eight votes to seven the proposal to mine polyhalite, used in agricultural fertilisers, including a mine shaft at Sneaton, accompanied by other infrastructure across the landscape to Teesside. A commercial opportunity to exploit for profit a resource in this way has driven a coach-and-horses through the principle of protecting the landscape for future generations - it will never be the same again.

Over many years the Association has sustained a forensic grip on the labyrinthine planning process to seek

The Founders of NYMA: Major Peter Walker, Derek Statham, Gerald McGuire and Don Tilley

to challenge the polyhalite mine project and subsequently to minimise the impact of this highly controversial scheme. Keeping informed and up to date in order to be heard and treated seriously is an unglamorous, daunting, but necessary task in the battle to try to 'hold the line' for the quality of the countryside into the future.

Even though the History Tree Plate was laid to commemorate a quarter century of NYMA, I would still have added its foundation into the series of memorable events mimicking the tree-ring growth on the Plate. Without the Association and organisations like it, who would there be to champion the cause?

So, these would be my nominations to add to the fascinating marks in time explored in this book. The 40 essays which follow, each by an author passionate for their subject, take us deeper into an extraordinary range of 'things' which the tree lived through, to engage us with the past, fire enquiring minds to discover more in the present, and most important of all to encourage us to look after the North York Moors National Park into the future.

Whenever I think of the tree and gaze at the space where it once stood, I am invariably left with the same wistful thought ….. 'it's a pity it had to be cut down'. But then I am lifted by a quote from Carolus Linnaeus, the 18th century Swedish botanist, physician and zoologist who defined the way we group and name plants: 'If a tree dies, plant another in its place'. And that's exactly what the National Park Authority did and what the North Yorkshire Moors Association continues to do when it has an appropriate opportunity.

So, as you enjoy these 'stepping stones' of the History Tree's life over the last two centuries, you might like to speculate a little on the future and the events which the replacement tree might live through. Have we any idea at all? After all, looking back I wonder whether the people who planted the History Tree could even remotely have begun to imagine what it was destined to live through, and the stories now recorded in this book.

Ian Carstairs, OBE
President: North Yorkshire Moors Association

All Saints Church, Brompton-by-Sawdon

William Wordsworth in 1798

The Nave, All Saints Church

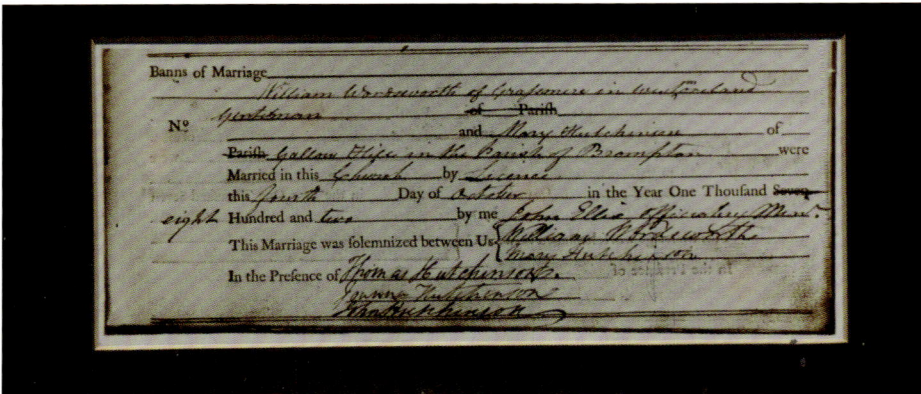

William Wordsworth and Mary Hutchinson's Marriage Banns, All Saints Church

The view from Sutton Bank

1802

The Marriage of William Wordsworth and Mary Hutchinson

Colin Speakman

William Wordsworth (1770-1850) was one of England's greatest poets. A lifelong walker and passionate lover of the countryside, especially his native Lake District, his ideas about the spiritual value of landscape and natural beauty still drive the worldwide concept of national parks.

Much of the emotional intensity and profundity of his greatest work came from the extraordinary relationships he had with two women - his sister Dorothy and his wife, Mary Hutchinson.

Twenty-one months younger than William and separated from her brother as a child after the death of their mother when she was seven, Dorothy lived first with an aunt in Halifax and, later, grandparents in Penrith until her early twenties. When re-united with William in 1794 they immediately formed a deep emotional bond which lasted for the rest of their lives. Dorothy, highly sensitive and intelligent, became both inspiration and muse to her gifted brother. Mary Hutchinson, an orphan whom Dorothy met when she was a pupil at a 'Dame School' in Penrith, became her lifelong friend.

Like many comfortably-off young men of his time, after graduating from Cambridge University William travelled in Britain and mainland Europe, a freedom denied his sister. He took walking tours in Italy and Switzerland and became obsessed with the radical ideas of the French Revolution. In 1791 he met and fell in love with Annette Villon, and, promising to marry her, fathered a child, Caroline. But the Reign of Terror in France forced him to return to England.

In 1795 William and Dorothy set up home together, first in Somerset, later in Grasmere. Mary, who now lived with her brothers and sister at Gallows Hill, near Brompton-by-Sawdon, was a frequent visitor.

With strong ties of love and friendship between the three, marriage between William and Mary seemed a natural step. It legitimised what was to be a platonic ménage-a-trois, in which sister, wife and poet shared a household.

In August 1802, when the international situation temporarily eased, William and Dorothy took what must have been a tearful, guilt-ridden trip to Calais to explain their forthcoming marriage and say farewell to Annette and Caroline, to whom William gave a small annuity. They then travelled to Mary's home at Gallows Hill for the wedding, arriving on 24th September.

Dorothy's journal gives a moving account of their wedding day on 4th October. The ceremony took place in Brompton's beautiful little church of All Saints, where a copy of the certificate can still be seen. Dorothy looked after the wedding ring, bought in France, keeping it safe on her own finger until the morning of the wedding, but was not present in church to witness the event. When the news reached her that the marriage had taken place she recalls: "I could stand it no longer and threw myself on the bed neither hearing nor seeing anything till Sara came upstairs to say, 'they are coming'".

Dorothy describes in detail the journey by carriage the threesome made 'straight after breakfast' from Brompton to Kirkbymoorside, where an official announcement was posted to the York Herald and time was spent deciphering gravestones, before continuing to Helmsley, where they spent the night. Descriptions of Helmsley Castle, Duncombe Park and Rievaulx Abbey follow. It was dusk when they reached Sutton Bank the next day and enjoyed views across Gormire Lake before arriving in Thirsk in darkness to find a bonfire in the Market Place to celebrate the local squire's birthday, and every bed taken at the inn. They had to continue to 'Leming Lane' to find a bed – at 11pm. The experience of the twilight ride down Sutton Bank inspired Wordsworth's sonnet "Dark and more dark the shades of evening fell".

It was a long, probably happy, marriage, despite the loss of William's brother John in 1805 in a shipwreck, and the early death of two of their five children. Dorothy suffered a major debilitating illness in 1829 and remained an invalid for the rest of her life, but she outlived her brother, only dying in 1855. Mary died a few years later, in 1859.

Cynics have suggested that after his marriage, Wordsworth's poetry never had the same intensity, radicalism and passion. But Dorothy and Mary undoubtedly made massive contributions to Wordworth's greatness as a poet – one his muse, the other his emotional anchor: a strange yet dynamic relationship which produced some of the greatest poetry in the English language. The relationship was at least formalised, if not celebrated, on the edge of the North Yorkshire Moors.

Find out more

All Saints Church, Brompton-by-Sawdon, has a copy of the marriage certificate of
William Wordsworth and Dorothy Hutchinson.

Wordsworth's life and contributions
https://wordsworth.org.uk/

'The Life of Wililam Wordsworth: A Critical Biography',
by John Worthen (Wiley & Sons, 2014)

1814

Louis Hunton, a Scientist who Changed the World

Mike Windle

Louis Hunton (1814-38) was one of the people who helped shape the modern world. His achievement in the science of geology is still used today and marks out his short life as one which deserves our gratitude and recognition.

Son of William Hunton, the Alum Agent at Loftus Alum Works, Louis was born at Hummersea House on the windswept Yorkshire coast in 1814, the eldest of nine children. He was christened Lewis but later changed his name to Louis out of respect and admiration for French scientific and social progress. He was educated in a school for local children and spent a great deal of time exploring the local quarries and their associated rocks and fossils, from which he managed to uncover a very useful fact, namely that fossils could be used to tell which rocks you were looking at.

The Lower Jurassic rocks on this part of the coast are approximately 180 to 200 million years old, give or take a few million years. Formed at the bottom of an ancient tropical sea that teemed with life, these rocks contain many different fossils including huge marine creatures such as Plesiosaurs and Ichthyosaurs. They also hold vast numbers of ammonites, squid-like sea creatures that evolved relatively rapidly and left their many different shells as markers of the passage of time.

It was known by the early 19th century that certain types of sedimentary rocks were formed in layers, piled one on top of the other, with the oldest normally at the bottom and the youngest at the top. But how to match these rock piles - or sequences - if they were geographically separated? This was the question which the young Louis was able to answer.

He studied rock sequences all over Cleveland, sampling and noting thousands of ammonite fossils from every sequence. By carefully identifying each type of ammonite, he was able to create a fossil sequence for each location and then match the sequences over many miles. He was meticulous in his collection and noted that only fossils still embedded in the rock face should be used: up to this point, geologists often collected their specimens from the debris at the foot of cliffs, so they were not always sure exactly where they came from within the rock sequence.

Louis' discovery is now known as biostratigraphy, the science of dating and correlating rock sequences, and is used by geologists all over the world for processes such as oil and gas prospecting, understanding global climate change, and learning about geological history.

At the age of 21, Louis wrote his only scientific article on his discovery, entitled: 'Remarks on a section of the Upper Lias and Marlstone of Yorkshire, showing the limited vertical range of the species of Ammonites and other Testacea, with their value as Geological Tests'. This was read to the Geological Society of London on 25th May 1836. We cannot be certain of its immediate effect on the geologists of the day but early the following year the influential geologist Charles Lyell addressed the Society and noted how useful Hunton's ideas would be for future geological correlation.

Sadly, Louis' work appears to have drifted into the shadows of history and was largely ignored for nearly two centuries, until 2014, when a project was launched to mark the 200th anniversary of the birth of this remarkable man and to commemorate his life and achievements. The project was led by the North East Yorkshire Geology Trust in partnership with a host of other supporters, and included a Service of Thanksgiving for the life and work of Lewis Hunton at St. Leonard's Church in Loftus, the church where he was baptised. In attendance were the Bishop of Whitby, the local Member of Parliament, four local mayors, Louis' biographer Dr Hugh Torrens, many local people - and of course several geologists!

Plaques were erected at his birthplace of Hummersea House and in the market square of Loftus, and an interpretation panel explaining his life and work was placed on the Cleveland Way within sight of his birthplace and next to the quarries where he discovered the secret value of ammonites.

Louis died tragically young at the age of only 23 years, in Nimes, France, of tuberculosis. In 1843 an ammonite - Tragophylloceras huntoni – was named after him, and in other ways too he left a lasting, practical and valuable legacy for the generations that succeeded him.

Find out more

Whitby Museum, Pannett Park, Whitby YO21 1RE,
https://whitbymuseum.org.uk/

Lewis Hunton Trail (Loftus) - self-guided walk leaflet available at
https://www.walkingloftusandthenorthyorkshirecoast.com/self-guided-walks

North East Yorkshire Geology Trust,
http://www.neyorksgeologytrust.com/pages/archive/lewishutton.html

Hunton's drawing of strata

Towards Boulby Cliffs from Cow Bar, Staithes

Ammonite fossil

Plaque: Hunton's House, Hummersea

Mammoth's tooth found in Kirkdale Cave

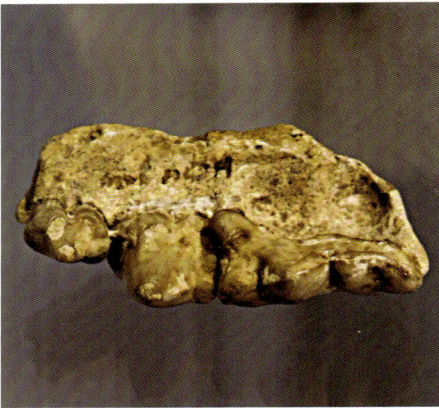

Part of a hyena's jaw found in Kirkdale Cave

Hodge Beck, Kirkdale, close to the cave

Kirkdale Cave

1821

Kirkdale's Contribution to the Theory of Evolution

Louise Mudd

The chance discovery of one of the most exciting finds of the early 19th century, for British geologists and palaeontologists at least, happened in the most unlikely of places. Hidden away in a secluded valley close to the market town of Kirkbymoorside, Kirkdale's only place of interest to locals and visitors alike up to that point had been the ancient Minster of St Gregory's, with its wonderfully inscribed Saxon sundial.

All this was to change in the summer of 1821, when workmen quarrying the limestone rock-face above Hodge Beck, just to the east of the church, exposed the entrance to a network of caves. Amongst the debris, locals began to uncover animal bones and teeth, which in turn attracted the attention of a young visitor from London. He went in search of the source of these fossils, and on arrival at the quarry discovered that a narrow entrance in the rock face had been revealed, about 40m above the riverbed of Hodge Beck.

After careful exploration he was greeted with an amazing haul of fragmented bones and teeth embedded in the cave floor, which was covered in thick blue clay. The visitor was told that a couple of years previously workmen had uncovered other bones, but they had been thrown into the river or simply added to the stone used in road surfaces: the site lay close to the old road between Kirkbymoorside and Helmsley.

Realising that this was something rather special, the young man headed back to London with his finds and soon word spread amongst the museums, reaching one of the most interesting characters in the field of geology: William Buckland.

Buckland was born in 1784 and developed a passion for fossils at an early age. He went on to study at Oxford University and was soon visiting various parts of England and Europe for his research into the expanding field of and paleontology. The news of the caves in Kirkdale was too good an opportunity to miss and Buckland came up to explore and record the findings in what proved to be a treasure trove of fossils.

With the permission of the neighbouring landowners, Mr. Salmond of York was hired to oversee the clearance of the caves. As word spread, more and more interested parties came to see for themselves what was happening. Not ones to miss a business opportunity, the local lads went in search of the previously discarded bones and teeth along Hodge Beck and sold them on to these academic visitors. One lucky lad sold a tusk from a bear for one and a half guineas – over £100 in today's money!

Once the bones were removed William Buckland began to catalogue and identify them, carefully documenting the condition of each one. In all he found 23 species of animal, many of which were not previously known to have lived so far north, such as hippopotamus, elephant and rhinoceros. Many of the bones were identified as hyena, which gave the cave the title of 'the Hyenas' Den'. He believed that the remains had been dragged into the caves by hyenas and showed that it had been their home for generations.

Buckland raised the notion that the bones could not have been washed up there from the tropics by the Biblical Flood, as others believed; he stated that these remains pre-dated the Flood and at best had been covered in mud by the deluge. At the time, these ideas challenged orthodox beliefs: Buckland himself was an ordained priest, the process of evolution had not yet been recognised,

and it was believed that all life had been created by God in its present form. His work in Kirkdale was regarded as the benchmark for how careful scientific research should be undertaken and made him the major figure in British geology, bringing him fame and recognition.

The cave network itself became an attraction, with Victorian and Edwardian gentlemen coming to explore the 90m system, with only the aid of candles to make their way through the tight passages, soaking in water and sticky mud. As the years passed the excitement died away, and today it would be easy to miss Kirkdale's caves, as the site has been bypassed by the new main road and most tourists again only come to see St Gregory's Minster. Local cavers, however, with their specialist equipment, continue to explore ever deeper into the limestone systems. Who knows if there are other secrets yet to be revealed in the Hyenas' Den?

Find out more

Ryedale Family History Group Research Room, Hovingham,
http://www.ryedalefamilyhistory.org/

Yorkshire Museum, Museum Gardens, York,
https://www.yorkshiremuseum.org.uk/

Kirkbymoorside History Group Archives,
http://kmshistory.btck.co.uk/

1827

Captain Cook's Monument

Carolyn Moore

Approximately eight miles west of where the 'History Tree' stood, at the National Park Moors Centre at Danby, is one of the most iconic landmarks in the North Yorkshire Moors. Erected in 1827, Captain Cook's Monument stands on Easby Moor at over a thousand feet above sea level and can be seen from many high points on the Moors and much of Teesside.

The monument was erected in memory of Captain James Cook (1728-1779), Fellow of the Royal Society, explorer, navigator, cartographer and captain in the Royal Navy. Cook was born and brought up in North Yorkshire. His birthplace of Marton is approximately six miles northwest of the monument, while his childhood home of Great Ayton can be seen about a mile to the northwest. Cook's formative years took him to the coast, first to Staithes as a grocer's apprentice, and then to Whitby as a Merchant Navy apprentice. The latter would eventually lead to a career with the Royal Navy, where Cook's worldwide explorations began and eventually brought about his demise at Kealakekua Bay in Hawaii.

Robert Campion, who funded the monument, was primarily a banker and also a sail-maker, with businesses in Whitby. He was also Lord of the Manor of Easby during the second quarter of the 19th century (although he became bankrupt in 1841). As explained in the Whitby Panorama and Monthly Chronicle of 1827, appeals for subscriptions for a monument were made from 1811; however, none were forthcoming. Campion's virtue and benevolence is extolled in the article as it tells how he erected the monument through his own expenditure. At the time of the appeal, a monument was intended for Eston Nab or Roseberry Topping, but it was eventually built on Campion's own Easby estate due to its elevation and the views it commanded towards Cleveland, Durham and the coast.

Campion laid the foundation stone on his birthday, 12th July 1827, witnessed by several onlookers including his own family and Captain Cook's nephew, Mr Fleck, who followed family tradition and became a master-mariner. The event is described as a celebration, with music playing and wine and spirits shared with the onlookers and workmen, whilst a ship's flag was hoisted. It is said that commemorative documents were placed within the foundations, including a portrait of Captain Cook and a glass plate etched with details of the event.

The monument was to be an obelisk 12 feet square at its base and was intended to stand 40 feet high, but when it was completed later that year on Cook's birthday, 27th October, it reached 51 feet tall. A further ceremony took place with an address by Campion followed by his son, John, scaling the scaffolding and placing a top-stone on the construction. The completion of the monument was celebrated in a similar fashion to the laying of the foundation stone three months earlier, with the addition of cannon being fired at the top of nearby Borough Green Woods, which echoed around the Moor.

An inscription, written on three cast-iron plates and sited on the west side of the monument, commemorated Captain Cook as "among the most celebrated and most admired of the benefactors of the human race" with the monument erected "as a token of respect for and admiration of the character and labours of this truly great man".

The original monument had a doorway and no railing. Unfortunately, it fell into disrepair towards the end of the 19th century. By 1894 an appeal was launched for its restoration and subscriptions were given, and by July 1895 the restoration had been completed. On 25th July the North East Daily Gazette describes how the restoration had saved the monument from further dilapidation caused by two lightning strikes which ripped off the door - leaving open access for sheep - and obliterated the inscription on the cast-iron plaques. The restoration included fitting the monument with a lightning conductor, covering the doorway and engraving an inscription on a granite slab paraphrasing the original. The then Lord of the Manor of Easby, Mr John James Emerson, gave permission for the restoration, provided stone from his own quarries and paid for a palisade to surround the base of the monument.

During the 120 or so years since its restoration, the monument has suffered serious damage just once, from a lightning strike in 1960, which split it virtually from from top to bottom due to corrosion of the lightning conductor. The monument was repaired relatively quickly and looks today much as it did in 1895.

Find out more

Roseberry Topping and Captain Cook's Monument Walking Trail,
https://www.nationaltrust.org.uk/roseberry-topping/trails/roseberry-topping-and-captain-cooks-monument

Captain Cook Birthplace Museum, Stewart Park, Marton-in-Cleveland, Middlesbrough TS7 8AT,
http://www.captcook-ne.co.uk/ccbm/

Captain Cook Memorial Museum, Grape Lane, Whitby YO22 4BA, http://www.cookmuseumwhitby.co.uk

Captain Cook Schoolroom Museum, 101 High Street, Great Ayton, Middlesbrough TS9 6NB,
http://www.captaincookschoolroommuseum.co.uk

Captain Cook & Staithes Heritage Centre, High Street, Staithes, Saltburn-by-the-Sea TS13 5BQ

Captain James Cook collection at the Whitby Museum, Pannett Park, Whitby YO21 3AG, https://whitbymuseum.org.uk/

Cook Monument, Easby Moor

Captain James Cook in 1775

Cook Schoolroom Musuem, Great Ayton

Replica of HM Barque 'Endeavour', Whitby

Scrimshaw: engraved whale's teeth

Scrimshaw engraving of a whaling ship

Captain William Scoresby, Senior

Replica of Scoresby's Crow's Nest

1829

William Scoresby Senior, Arctic Whaler

Fiona Barnard

Born in 1760, William Scoresby grew up on his father's modest farm at Nutholm, near Cropton, on the southern edge of the North Yorkshire Moors. He left school aged nine to help on the farm but in 1780 sailed as an apprentice on the Jane, initially from Whitby to the Baltic. Scoresby left the Jane in London in 1781 to join the Speedwell, which was carrying stores to besieged Gibraltar during the 'Great Siege' by France and Spain (it lasted almost four years). The Speedwell was captured by the Spanish and the crew imprisoned, but Scoresby and a companion escaped to Cádiz, where they stowed away on an English ship. Scoresby returned to the farm and in 1783 married Lady Mary Smith, daughter of a Cropton farmer (the name 'Lady' was commonly given to girls born on 'Lady Day' – 25 March). They had eleven children, five of whom survived infancy. Following his wife's death in 1819 he married Hannah Seaton of Hull.

In 1785 Scoresby joined the Henrietta, sailing to the Greenland whale fishery. The captain recognised his talents and he rose rapidly through the ranks to become chief harpooner by 1790. When the captain retired in 1791 he recommended that Scoresby should replace him. Although on his first voyage as master no whales were caught, Scoresby was given another chance and picked his own crew for 1792: he returned to port with the produce of 18 whales, the largest catch by a Whitby ship up to that point. In the following years Henrietta continued making record catches and Scoresby became famous for his skill. In 1798 he accepted a lucrative contract to command the Dundee of London.

In 1800, when Scoresby called in at Whitby to see his family on his way north, his 10-year-old son William hid on the ship when it was time to leave. His father sailed with him on board rather than miss the tide, putting him ashore to be looked after in the Shetlands; but the boy quickly escaped and returned to the ship, where he stayed for the whole voyage.

After four successful years with the Dundee, Scoresby returned to Whitby to take command and an eighth share of the new ship Resolution, with his son William now an apprentice. In May 1806, with 16-year-old William already first mate, the Resolution was in ice off Svalbard when Scoresby detected a shadow on the horizon which he deduced was open water. He forced the ship through and after five days reached an immense open sea which he explored northward to the latitude of 81° 30', the furthest north any ship had sailed. That voyage yielded 24 whales, 1 narwhal, 2 seals, 2 walruses and 2 polar bears. Later in 1806 William Junior began studying at Edinburgh University, re-joining the Resolution each spring.

In 1807 Scoresby introduced the crow's nest, a barrel-like structure atop the main mast, protecting the watching officer from the weather as he navigated.

Scoresby passed command of Resolution to William Junior when he turned 21 in 1810. He himself moved to Scotland to take a share in the Greenock Whale-Fishing Company and command their ship John. After the 1814 season he resigned his command to his daughter Arabella's husband, Thomas Jackson. In 1816 he took command of Mars of Whitby and a year later bought Fame, hoping - in vain - that the Admiralty would accept his proposals for an expedition to Greenland. He continued whaling with diminishing success until 1823, when Fame was destroyed by fire in the Orkneys on her outward voyage.

Scoresby was by then 63 and whales were becoming scarce, so he retired to enjoy his considerable fortune. In retirement he published proposals for improvements to Whitby's roads and harbour and for employment of the poor. He died in 1829. Captain William Scoresby was a large, energetic man with great strength and stamina. His acute observations added to his exceptional abilities as navigator and seaman. He made many practical improvements in the whaling industry, including changes to the rigging and ballasting of his ships which made them safer and more manoeuvrable, but his major innovation was the widely adopted crow's nest, which gave navigators some shelter and security. His record for sailing the furthest north of any ship stood for several decades.

He insisted that his children were educated. He was a deeply religious man, a conviction which he passed to his son William (1789-1857), who became a distinguished whaler, scientist, explorer and clergyman, and wrote the definitive book on Greenland whaling.

Find out more

Whitby Museum, Pannett Park, Whitby YO21 3AG,
https://whitbymuseum.org.uk/
(Archive material may be viewed by appointment)

'My Father' by The Rev. William Scoresby,
(1851, reprinted by Caedmon Press 1978)

1836

The Whitby and Pickering Railway

Andrew Scott

Whitby's early 19th century wealth came via the sea, since land communications across the boggy moorland had always been difficult. Then in 1825 Whitby's merchants saw the impact of railways on their Stockton rivals. George Stephenson had built the Stockton & Darlington Railway for local coal-mine owners, transforming the wealth of the Tees Valley by bringing coal to the coast quickly and cheaply. Whitby's future clearly depended on better links, but what sort - and where should they go? Plans for a canal were soon discarded and the advice of George Stephenson was sought.

Stephenson explored options, surveyed a route and, after parliamentary approval, supervised construction; he was a busy man, however, and the exact survey of the chosen route to Pickering was prepared by Fred Swanwick, one of his young assistants, who subsequently became a significant railway builder himself.

After opening first from Whitby to 'Tunnel' (Grosmont), services ran to Pickering from May 1836. But the Whitby & Pickering line had necessarily been built cheaply. Structures could not carry the heavy steam locomotives used elsewhere; horses provided the motive power, while passengers travelled in mail coaches on railway wheels. To lift carriages up onto Goathland's moors whilst avoiding gradients too challenging for horses, a mile-long inclined plane was installed where the weight of descending loads lifted ascending trains via a connecting rope. The descending load had always, of course, to be heavier than that ascending. This was achieved by travelling water tanks, filled from a reservoir at the top of the incline and emptied at the bottom.

Whilst the new railway worked, its low capacity, limited market and high costs brought commercial failure. In 1845 it was saved by George Hudson, who was busy shaping a network of railways across the north, including a line from York to Scarborough. Hudson bought the railway at a bargain price and rebuilt it to higher standards, laying double track over new, stronger, bridges. A bigger tunnel was made at Grosmont and new stations were built. Crucially, the line was extended to join the new Scarborough to York line. From 1847, locomotives pulled full-sized trains from Whitby to Pickering and onwards to York and even London. But the route still included Goathland incline, and for another eighteen years, trains were raised or lowered on the end of a rope between Goathland and Beckhole. Charles Dickens noted the system's curiosity in letters written to Wilkie Collins at Whitby in 1861. The system prevented extra holiday trains being run to Whitby and it was feared by the public: in 1864, two people were killed in a runaway after the rope broke. Change was needed.

By 1860 most of northeast England's railways had been consolidated into a huge regional monopoly, the North Eastern Railway. In 1865 the NER improved lines serving Whitby and the Moors, not least because of the increasing ironstone trade from the area. Eventually, a new line was built from Grosmont to Moorgates, above Goathland, bypassing the rope incline but requiring heavy engineering works as the line climbed a three-mile 1:50 gradient – near the limit for contemporary locomotives. The Esk Valley's railway network was now complete.

A century later Dr Richard Beeching was charged with reshaping the railway for the motor age. Much argument resulted over which of Whitby's routes should be maintained, and eventually the Esk Valley villages' need for school transport prevailed over Whitby's connection to York and the south. The line from Grosmont Junction through Pickering to Malton closed in 1965.

Public reaction to the Beeching closures was immense and the loss of services south from Whitby was no exception. In June 1967 local people launched a Preservation Society. Volunteers came together, funds were raised and public support was garnered. In 1968, once a deposit was paid for the railway's land between Grosmont and Pickering and the track as far as the summit at Ellerbeck, work started to resuscitate the railway. Later, the County Council helped the Society buy the rest of the track through a mortgage.

After a huge amount of hard work to provide the locomotives, carriages and maintenance facilities which a free-standing line needs, the North Yorkshire Moors Railway opened in May 1973. It has never received regular public funding but thanks to the efforts of staff and volunteers it has become the world's busiest steam heritage railway, with over 300,000 passengers annually. Uniquely, it has gained powers to run trains onwards from Grosmont over Network Rail's line into Whitby, thus reinstating the original service of the Whitby & Pickering Railway.

Find out more

North York Moors Historical Railway,
https://www.nymr.co.uk/

'A History of the Whitby and Pickering Railway'
by G.W.J. Potter (1906, reprinted 1969)

North Yorkshire Moors Railway: Levisham

William Smith in 1837

Smith's geological map

The dome in the Rotunda Museum, Scarborough

An ammonite, Rotunda Museum

Rotunda Museum

1839

William Smith, the Father of English Geology

Alan Staniforth

William Smith caught a chill while on his way to a meeting of the British Association in Birmingham and died a few days later on 28th August 1839. He was buried at Northampton. In his three score years and ten he had risen from obscurity to become one of the most famous men of his time.

Smith was born on 23rd March 1769 at the village of Churchill in Oxfordshire. His parents were of humble origin, working the rich land in the neighbourhood. Like many of his day, William had little opportunity for any formal education, but he had a good retentive memory and was a keen and accurate observer, attributes he would put to good use in later life.

Smith's mother remarried when his father died in 1777 and at eighteen William became an assistant to Edward Webb of Stow-on-the-Wold. It was working under Webb, first on land enclosure maps and later on the survey of canal lines, that gave Smith his first opportunity to enlarge and apply his embryonic geological knowledge.

With the passing of the Canal Bill in 1794, Smith - still in his early twenties - found an increasing market for his specialist and largely self-taught skills. On a circuitous 900-mile journey to the north of England to survey canal lines and mining operations in Yorkshire, he was able to extend his observations and was soon in a position to relate the rocks in the north to those he had already studied in the south of England.

Following this extensive tour, Smith was engaged for six years in superintending the works on the Somerset Coal Canal. With his increasing geological experience, he was able to advise the contractors on the best method of work. It was also now that he collected "extraneous fossils" and realised "that each stratum contained organised fossils peculiar to itself, and might, in cases otherwise doubtful, be recognised and discriminated from others like it, but in a different part of the series, by examination of them".

This is the first reference to one of Smith's great contributions to the science of geology, that characteristic fossils can be used to correlate strata across the country. Using this technique, he was able to produce, in 1815, the first large-scale geological map of Britain - or indeed of any country in the world. Printed on fifteen sheets to a scale of five miles to the inch and measuring 6 feet by 8 feet 6 inches, an original copy hangs in the Yorkshire Museum. A work of art in its own right, a comparison with a modern map shows only minor differences and serves to illustrate the accuracy of Smith's observations and recording. This was the dawn of stratigraphical geology and geological cartography. To celebrate the 200th anniversary of the publication of this great map a mosaic was created in the Museum Gardens in York which depicts the Yorkshire section in coloured pebbles.

Sadly, the publication of his map was overshadowed by other matters, mainly financial. Smith was forced to sell his extensive geological collection to the British Museum in order to pay his debts, but still spent ten weeks in the debtors' prison. In 1819 he gave up his London house and moved to the north of England, eventually settling in Scarborough.

It was at this time that the Scarborough Philosophical Society was formed and the members decided to build a museum, the Rotunda, at the Aquarium Top. The museum's circular design is credited to Smith, who suggested that geological specimens could be better displayed in their correct ascending order in a round building. A painted geological section of the Yorkshire coast based on the work of Smith's nephew John Phillips can still be seen around the inner frieze of the dome. Although later used to display archaeological and local history material, the museum was extended and restored to its original purpose in 2008 and dedicated to William Smith. Between 1828 and 1834 Smith lived at Hackness, where he was employed as land steward to Sir John Johnson of Hackness Hall.

At last, late in life, recognition began to appear. He was given an Honorary Doctorate by Dublin University, and the Geological Society, which had shunned him in earlier years, now made amends and welcomed him into their midst. In 1831 he was awarded the first Wollaston Medal for services to geology. It was the President of the Society who, in presenting the medal, dubbed Smith 'The Father of English Geology'.

Find out more

'The Map that Changed the World' by Simon Winchester (2001)

'Strata' by John L Morton (2004)

Rotunda Museum, Vernon Road, Scarborough YO11 2PS,
http://www.scarboroughmuseumstrust.com/rotunda-museum/

Yorkshire Museum, Museum Gardens, York – including The Mosaic Map, the Yorkshire section of Smith's map,
https://www.yorkmuseumgardens.org.uk/the-mosaic-map/

Interactive website on William Smith's maps
http://www.strata-smith.com/

1844

Charles Dickens visits Mulgrave Castle

Ann Glass

In April 1844 Charles Dickens, one of the most eminent authors and social reformers of his time, visited Mulgrave Castle at Lythe, near Whitby. Mulgrave Castle was the home of Lord and Lady Normanby. Lord Normanby was a politician and author. So, what was it that brought Dickens to Mulgrave Castle, and how did he respond to the North Yorkshire Moors?

Dickens was born in 1812, the son of John Dickens. His father lived beyond his means and when Dickens was twelve years old was imprisoned in Marshalsea debtors' gaol. Charles was removed from school and sent to work in a factory to support his family. His early experiences shaped his views and provided rich fodder for his writing, including a concern to expose the plight of the working classes. His articles and sketches developed into part fact, part fictional stories, and were eagerly awaited by scholars, politicians and workers alike. The articles, and subsequently his novels, brought the ills of the era to public attention in an accessible form for the first time. His personality contributed to his success and he became something of a celebrity.

Dickens treated his novels as a springboard for debate about moral and social reform. He was extremely sociable and influenced his friends and acquaintances in informal and formal gatherings. His celebrity status and substantial income allowed him to travel widely, to indulge his passion for theatrical performance and develop into an accomplished networker.

Whilst travelling his pen was never idle. His lively imagination and his powers of observation conjured up scenes and characters such as Squeers in 'Nicholas Nickleby', thought to be based on a headmaster he visited during a trip to Yorkshire to look into the condition of schools. In the early 1830's Dickens met Charles Smithson, whose family law practice was based in Malton, and they became friends. When Smithson's elder brother and father died he returned to Malton from London to run the family practice in Chancery Lane: it is suggested that the building became the model for Scrooge's counting-house in 'A Christmas Carol'. In 1843 Dickens spent three weeks visiting Charles Smithson, enjoying his hospitality and the North Yorkshire countryside.

He enjoyed the wide open spaces and was clearly delighted: "all day long I cantered over such moss and turf that the horse's feet scarcely made a sound upon it".

As Dickens' popularity grew he mixed with the most influential figures of his time, including Lord Normanby, a Whig politician with liberal views who was Home Secretary from 1839 to 1841. Dickens and Lord Normanby shared a love of oratory and amateur theatricals, which was also a passion of Lord Normanby's son, Lord Mulgrave; Dickens had crossed the Atlantic five years earlier and performed with him in Montreal. Lord Normanby was also a novelist.

In April 1844 Charles Smithson died prematurely, and following the funeral in Malton Dickens was invited to stay at Mulgrave Castle by Lord Normanby, who spent several days walking and riding with him and showing him the district, including a visit to Whitby Abbey and to the fishing village of Staithes. They lunched at 'The White Horse and Griffin' in Church Street, Whitby. Dickens also travelled on the - then horse-drawn - railway to Pickering (now the North Yorkshire Moors Railway).

He was delighted with his stay at Mulgrave Castle and is reputed to have been so ecstatic about its situation and the views from the newly designed terraced lawn, 'The Quarter Deck', that he danced with joy. He expressed himself to be surprised at so many beauty spots in such close proximity.

In July 1844, shortly after his return to London from Yorkshire, Dickens hosted a dinner to celebrate the publication of 'Martin Chuzzlewit' which was presided over by Lord Normanby and attended by many notables. In 1847 Dickens was to dedicate 'Dombey and Son' to the Marchioness of Normanby.

In later years Dickens fondly remembered his trips to North Yorkshire. He wrote to his great friend, the novelist Wilkie Collins, and referred to the Pickering to Whitby railway thus: "that curious rail-road on the Whitby Moor - you were balanced against a counter-weight of water and that you did it like Blondin" (Blondin being a renowned French tightrope walker of the time). It was clearly a striking memory amongst many.

The particular combination of Dickens' personal attributes, interests and consequent friendships begin to explain how he came to visit to Mulgrave Castle in 1844 and how that contributed to his appreciation of North Yorkshire.

Find out more

Dickens Society, Malton
http://dickenssocietymalton.co.uk/

Mulgrave Castle
https://historicengland.org.uk/listing/the-list/list-entry/1001065

The White Horse and Griffin, Church Street, Whitby

Charles Dickens in 1867-68

Inspiration for Scrooge's Counting House

Mulgrave Castle, Lythe

Road-sign carved by John Castillo, Stonegate

The River Esk, Lealholm

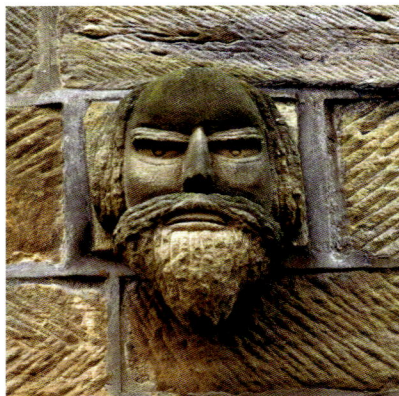

Modern tribute to Castillo, Lealholm

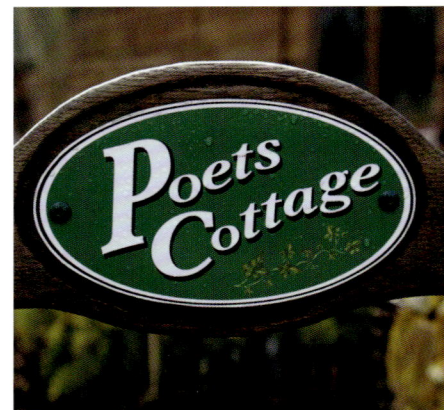

Sign near Castillo's former home, Lealholm

1845

Poet John Castillo, Bard of the Dales

Jane Ellis

John Castillo, 'The Bard of the Dales', was a stonemason and poet, born in 1792 in Ireland; his father, Patrick Castlehowe, was an itinerant Irish labourer who had come to Yorkshire seeking work, and married Mary Bonas (or Boanas) of Glaisdale, at Danby Church. When John was very young his father moved the family back to Mary's home valley, obtaining work at the paper mill in Lealholm and making a home at what would later be known as Poet's Cottage, though eventually Patrick left them to seek a better living, never to return. So bad was the family's poverty that when the young John began work in 1805 as a servant to a Lincolnshire gentleman, he needed Parish Relief to provide him with suitable clothes.

Within two years John was back in Eskdale, becoming a skilful journeyman mason and having a hand in building many properties and farm buildings which stand to this day. One example of his work, still admired by passers-by, is the useful and decorative direction sign at Stonegate which shows the mileages to Castleton and to Whitby, with carved pointing fingers. He left his mark on 'The Dancing Stone' in Danby Dale, which has his name carved into it with the later addition 'Neu hees dead' ('Now he's dead').

Having abandoned his family's Catholic faith John became a fervent Methodist, travelling extensively and preaching. He was a natural storyteller and wrote hundreds of poems and songs, many in broad local dialect; in fact, he was often composing when quietly working. He would commit his ideas to paper and stuff these into his hat, which was often full by the end of the day. Some of his most enduring legacies though are the carved heads with which he adorned properties throughout the area, using his remarkable talent for creating recognisable likenesses of people he knew. Along with three heads on Lealholm chapel, which he helped to build in 1839, he created that of a young Queen Victoria at her coronation.

His work was published in 'Old Isaac, The Steeplechase and Other Poems' (1843) and 'The Bard of the Dales or Poems of Miscellaneous Pieces' (1850), both packed with poetry inspired by local people and events. The latter contained a glossary of the dialect words he used, useful for enabling the general reader to understand what can appear to be a foreign language!

Castillo died in 1845 and was interred in the Wesleyan Chapel burial ground at Pickering, where he had settled in later life. His poem 'Lines on Leaving Fryup, in Search of Work' demonstrates how it pained him to leave his beloved Eskdale, but we can see that he faced this major change with his customary optimism and cheerfulness:

I'm sorry, Fryup! thee to leave,
But thou deniest what I crave,
Though I have ask'd with tears!
Oft I have drunk at thy pure rills,
And labour'd 'mongst thy
moorland hills,
For many toilsome years!

Here, with each morning's
early dawn,
I lov'd to walk the flowery lawn,
To hear thy warblers sing!
Or when at eve their
songs were mute,
I've sooth'd my fancy with my flute,
And made thy woodlands ring!

I've seen thy mountains
clad with snow,

While shelter'd in the vale below,
'Midst hospitable friends!
For all their kindnesses to me,
May Heav'n bless every family,
And make them full amends!

But trade is now so dull and dead,
A man can hardly earn his bread,
In winter's frost and snow:
So I must take my staff in hand,
And travel to some distant land,
Till here more plenty grow!

I grieve to leave the Sunday School,
Where I with gratitude of soul,
Have taught with great delight,
The youthful, rising sons of men,
To steer safe past the gulf of sin,
By glorious gospel light.

For quicksands and contrary winds,
And enemies as well as friends,
I still expect to find:
There is a Friend who lives above,
To all who do His precepts love,
He proves both true and kind!

To Him I will address my prayer;
My little bark unto His care,
With confidence I'll trust!
A steady course, O may I steer,
He'll land me safe at last!

His headstone has survived and is preserved in a private garden. Although he roamed the moors and dales two centuries ago, we still have our reminders of this larger-than-life character.

Find out more

'Poems in the Yorkshire Dialect' by John Castillo, edited by George Markham Tweddell (1878), download or read online at
https://huddersfield.exposed/wiki/Poems_in_the_North_Yorkshire_Dialect
_by_John_Castillo_(1878)_edited_by_George_Markham_Tweddell

The Dancing Stone in Danby Dale, above Botton, grid ref. NZ 701048.

Castillo's signpost with the pointing hand is on a roadside building at Stonegate, near Lealholm.

The Reverend John Atkinson: 'Forty Years in a Moorland Parish'

Louise Mudd

Anyone with an interest in the North Yorkshire Moors and its traditions and folklore must read this fascinating record of life in and around the village of Danby, published in 1891 by its rector, the Revd. John Christopher Atkinson. The book is remarkable in its depth and detail of a way of life which was already fading fast when he arrived in this remote corner of the world in the mid-19th century.

Born in Essex in 1814, John Atkinson came from a strongly religious and academic family. He studied at St John's College, Cambridge, and was ordained in 1841. Revd. Atkinson's association with North Yorkshire came through obtaining a position in 1847 as Lord Dawnay's Chaplain. In 1850 he was presented with the living at Danby, shortly after his marriage to Jane Hill. The couple arrived to find an insular and neglected parish, and so Revd. Atkinson set about establishing himself and gaining the trust of the locals.

For a man with a passion for nature and history, the remote parish provided a fascinating wealth of tales, legends and details. Alongside the challenges of daily ministry, he fathered eight children, but suffered the loss of his wife Jane in 1860. Having remarried in 1862, to Georgina Mary Slade, he had five more children but again was widowed before marrying for a third and final time in 1884, to Helen Georgina Brown. He wrote a series of children's books and on ornithology and birds' eggs, but it his work on local history for which he is best remembered.

'Forty Years in a Moorland Parish' is a collection of topics covering everything from the style of houses to local customs. Some of the most fascinating chapters cover local legends. The folklore Atkinson gathered revealed how ingrained some of the old beliefs still were at all levels of society. There are tales of hobs (small household spirits) and fairies as well as the more usual superstitions relating, for instance, to magpies.

The material had to be gathered slowly as many people were initially reluctant to open up, for fear of mockery. Quite often the source of the tales was deliberately left vague to protect people's identity.

Atkinson recognised that the old customs were fading fast as the modern world encroached. The very rural nature of his parish meant that he could reach further back in time than would have been possible in a town or more industrial area. The nature of farming was beginning to change with small family farms merging into larger ones, with traditions merging with them. Years of officiating at baptisms, marriages and funerals also gave Revd. Atkinson unique insights, and he also found the time to explore the connections in tradition between North Yorkshire and Scandinavia.

One such tradition involved the humble honey bee. A local practice was to put bee-hives into mourning after the death of their owner. Often the widow would go to the hive and inform the bees that one master was dead and tell them who their new master was, whilst offering them food and drink. A theory advanced by Atkinson was that this tradition stemmed from Scandinavian beliefs in Valhalla and the fear of restless spirits returning from beyond the grave. By telling the bees who they were to obey and providing them with sustenance, they would ignore any returning spirits seeking to steal them away.

Many dialect words and phrases were also disappearing as schooling and ways of speaking became standardized. Younger generations did not always continue to use the distinctive style of describing people: "he's that thin, he's lakh a ha'porth o' soap after a lang day's weshing" is so much more evocative than "he's rather thin"!

Atkinson was not just an observer: he also enjoyed getting involved in the physical discoveries made in the area. His writing on barrows and earthworks reveal how he would often be invited to excavations along with other academics and passionate volunteers. He opened up around 90 barrows on Westerdale Moor and described the stone axe-heads found on Skelton moors.

'Forty Years in a Moorland Parish' was a huge success. It has been reprinted several times since 1891 and remains a key reference source for local historians and academics.

In the same year as his book was published, Revd. Atkinson was installed as Canon of York Minster, although he remained in his beloved Danby until his death in March 1900. He left behind an invaluable legacy, for which we historians are truly grateful.

Find out more

'Forty Years in a Moorland Parish:
Reminiscences and Researches in Danby in Cleveland' by J.C. Atkinson (1891, facsimile edition 2006)

Atkinson's entry in an index of British naturalists:
http://www.natstand.org.uk/time/AtkinsonJCtime.htm

The Revd. John Atkinson in his study

St. Hilda's Church, Danby Dale

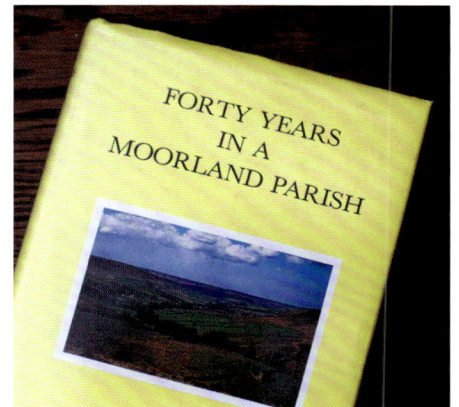

FORTY YEARS
IN A
MOORLAND PARISH

Atkinson's renowned book

Revd. John Atkinson's grave, St Hilda's, Danby

Sir George Cayley

Replica of Cayley's glider

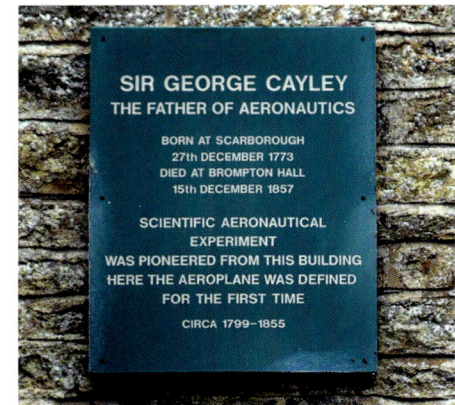

SIR GEORGE CAYLEY
THE FATHER OF AERONAUTICS

BORN AT SCARBOROUGH
27th DECEMBER 1773
DIED AT BROMPTON HALL
15th DECEMBER 1857

SCIENTIFIC AERONAUTICAL
EXPERIMENT
WAS PIONEERED FROM THIS BUILDING
HERE THE AEROPLANE WAS DEFINED
FOR THE FIRST TIME

CIRCA 1799–1855

Sir George Cayley's workshop, Brompton Hall

Plaque, Brompton Hall

1853

Sir George Cayley, The Father of Aeronautics

Albert Elliot

In 1853, Brompton Dale, near Scarborough, witnessed the world's first successful manned aeroplane flight. The flying machine was the invention of Sir George Cayley, of Brompton Hall, west of Scarborough. The hapless lone passenger or 'pilot' was Caley's harassed coachman, who became - reluctantly - the first person ever to take to the sky in a heavier-than-air machine. The wooden-framed aeroplane, manned (although not piloted), was towed off by a galloping horse and flew for about 275 metres across to the far side of the Dale, carrying the terrified coachman. Cayley's 10-year-old granddaughter, Dora, was an eyewitness on this momentous occasion and left us a brief but fascinating first-hand account of the glider's flight:

"Of course everyone was out on the high east side [of Brompton Dale] and saw the start from close to. The coachman went in the machine and landed on the west side at about the same level. I think it came down rather a shorter distance than expected. The coachman got himself clear, and when the watchers had got across, he shouted, "Please, Sir George, I wish to give notice, I was hired to drive, and not to fly". That's all I recollect. The machine was put high away in the barn and I used to sit and hide in it (from Governess) when so inspired."

Little did the coerced coachman, nor the amazed observers in attendance on that day, realise the great historic significance and future impact of that first flight. Cayley's incredible feat would change the world forever and lead directly to the commonplace modern-day phenomenon of air travel: thousands of jet aeroplanes daily crisscrossing the world's skyways carrying millions of passengers to far-flung corners of the globe. All this aerial activity can be traced directly back to that singular event in 1853 when Sir George's 'new flyer' lifted into the air and 'sailed away in graceful and stable flight' a relatively short distance across Brompton Dale. Mankind was airborne!

As well as being the creator of the first aeroplane, Sir George Cayley – also deservedly given the epithet 'A Yorkshire Genius' - was the source of many other ideas and inventions, such as the spoked tension wheel (forerunner of the modern bicycle wheel), a 'helicopter', artillery shells, land drainage systems, a continuous track mechanism (forerunner of the systems later used for the military tank and caterpillar tractor), an amphibious vehicle, agricultural machinery, fire safety curtains in theatres, train buffers, seat-belts for passengers, a self-righting lifeboat, a new type of pen nib, a new kind of parachute, an internal combustion engine (using gunpowder as the motive power!), and an anemometer for measuring the speed of air currents in mines. His 1809 design sketch for the aerofoil section of a wing, said to be based on the shape of a trout, is almost identical to the low-resistance aerofoil sections of today.

Such was his fecund mind that when the son of one of his tenants lost his right hand, Sir George set about designing and making him an artificial one. Dr Roget (famed for his Thesaurus and friend of the inventor) persuaded Cayley to present George Douseland, the recipient of the prosthetic hand, at a reception which Prince Albert attended. Douseland was said to have received a royal handshake from Queen Victoria's husband who clasped his iron hand when it was offered. Remarkably, the false hand weighed only one pound eleven ounces, only six ounces more than his natural hand.

A compassionate man, Sir George was an enlightened and benevolent landlord. He was the first to instigate a cottage allotment scheme for his tenants, who were each allowed a generous half-acre plot on which to grow food and produce for their families. A humanitarian, he always showed concern for the welfare of his fellow men. Ahead of his time, he suggested safety legislation that pre-dated by far the plethora of Health & Safety legislation that was to come.

Sir George Cayley was a truly remarkable Yorkshireman - undoubtedly a genius - who is only in recent times receiving the full credit that his many achievements deserve. Amongst the earliest to give him due recognition were a group of people much keener than his coachman to take to the skies. In 1975 they founded the 'Sir George Cayley Sailwing Club' for those who partake of the 'joy of free flight' by hang-gliding and paragliding in Yorkshire.

Find out more

The Yorkshire Air Museum, Elvington, York YO41 4AU,
has a life-size replica of the original sailplane,
http://yorkshireairmuseum.org/

The Royal Aeronautical Society has articles, a podcast and digitised copies of Sir George Cayley's notebooks,
https://www.aerosociety.com/

1853

Henry Cooper, the Scugdale Giant – a Tall Tale

Albert Elliot

Giants have featured in lore and legend since time immemorial. Perhaps the most famous of all comes from the Bible in the form of Goliath, who was slain by the diminutive shepherd boy, David, with his deadly slingshot. More local to our own area is the legend of Wade, a colossal giant said to be over 100 feet tall! Wade is reputed to have single-handedly built the road over Wheeldale Moor (the Roman road called Wade's Causeway) to make it easier for his wife Bell to tend to and milk her giant cow, which she kept on the moors. Wade is also said to have inadvertently created the Hole of Horcum ('The Devil's Punchbowl') when in anger he scooped up a handful of earth and flung it across the moor at his wife. Fortunately he missed his target, with the result that the prominent conical hill to the east of the 'giant's crater', known as Blakey Topping, came into being at the very spot where Wade's 'handful' fell to the ground.

However, many stories of giants are based on actual facts and known real people. From the North Yorkshire Moors comes the fascinating tale of the Scugdale Giant. Henry Alexander Cooper (known as Harry) was born 1853, although accounts of his place of birth differ. Some claim he was born at Swainby and as a boy worked on a farm in nearby Scugdale. Perhaps a more trustworthy account states that he was born at Robinson's Yard, Commercial Street, Norton, near Malton, although the Cooper family were known to have moved about the country in search of employment.

Following a few years' schooling, Henry was sent into farm service at High House, Scugdale. He then moved to Rosedale, becoming a mineworker in the thriving ironstone industry. It was there that he suffered a bout of illness that would dramatically change his future life. At the age of 16 he was struck down by a severe fever that kept him confined to bed for 13 weeks. During this period Henry is said to have undergone an abnormal growth spurt in which he grew by an astonishing 16 inches. He had become a giant, standing at 8 feet 6 inches tall - although this is perhaps an exaggeration as it would mean that he was already over 7 feet tall at the age of 16, prior to his illness. He was said to have hands 13 inches long and feet measuring an enormous 17 inches. Henry Cooper had become the tallest known man in Britain and a true Goliath.

He moved again to North Skelton in East Cleveland to work at Fogga, the local ironstone mine. The work in the pit must have been excruciatingly difficult for a man with such a huge frame. Henry was a popular character in East Cleveland and became affectionately known as Long Harry. He was a quiet, inoffensive man, although extremely strong. Tales of his unique stature and immense strength were legion.

One day a passing fairground owner spotted Henry and invited him to join his travelling circus: and so began his journey to fame as the world's tallest man. He toured England, visiting many cities and towns. While on tour Henry attracted the attention of Mr P. T. Barnum, who made an offer that Henry duly accepted. He was billed as 'The Yorkshire Giant - Tallest Man in the World', and travelled across America with Barnum's Travelling Show, the 'largest show on earth': quite an adventure for a humble farm boy from Scugdale.

While in the United States, Henry met and married a tall American lady with whom he had children. In 1886 an announcement was placed in the New York Clipper trade magazine stating that Minnie Cooper, Henry's wife, had given birth to a baby weighing 17 pounds. Henry was to tour America and Canada for the rest of his life with various circuses and sideshows.

In 1889, while making his way to Edmonton, Canada, Henry fell ill and had to be left behind in Calgary. Despite receiving medical attention, he passed away quietly at the age of 46. He died surrounded by colleagues of the local Oddfellows, a worldwide fraternal society of which Henry was a member. The Calgary Oddfellows arranged Henry's funeral, carried out the burial duties and assisted at the ceremony. His remains were interred at the Calgary Union Cemetery. At the time of his burial no gravestone was erected but in 2004 surviving members of Henry's family organised the erection of an inscribed memorial headstone.

Find out more

Information on Henry Cooper:
http://www.thetallestman.com/henryalexandercooper.htm

Story of P.T. Barnum and his circus:
http://www.historyextra.com/period/victorian/circus-sensation-pt-barnums-greatest-wheezes/

Henry Cooper

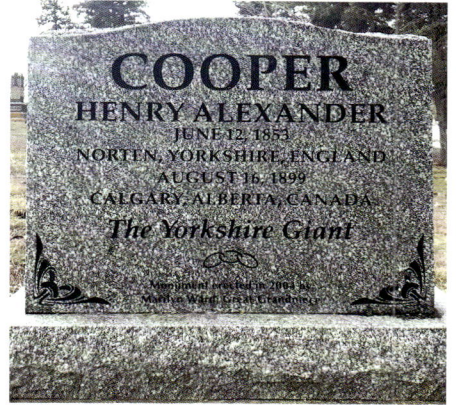

Henry Cooper's gravestone, Calgary

The road to the dale end, Scugdale

Scugdale

The White Horse, Kilburn

The Kilburn White Horse

Shirley Learoyd

The White Horse carved into the steep south-west facing slope of Roulston Scar, just above the village of Kilburn, in the North York Moors National Park, is the largest hill figure in the country and the only one in Yorkshire. Its dimensions are 97m long and 67m high, covering just over an acre of ground, and it is visible from many miles away.

The huge Horse was the idea of a local man, Thomas Taylor, a native of Kilburn, who in his travels around the country had seen other such horses cut into hillsides in Wiltshire and elsewhere. He wrote to his friend John Hodgson, the Kilburn village school-master, suggesting its siting, with a drawing by the artist Harrison Weir of an outline of the proposed carving. John Hodgson, who was also a land surveyor, took up this suggestion and set about surveying the area. Some of his pupils helped to lay out the outline of the Horse by driving stakes into the ground, and then volunteers from the village cleared the site to expose the greyish sandstone rock. Unlike the hill figures further south which were carved into chalk and thus naturally white, the Kilburn carving had to be whitened by using 6 tons of limestone chippings. It was finished in November 1857 – an event celebrated by the villagers with a huge party at which 100 gallons of beer were reportedly consumed and two bullocks were roasted.

The Horse is visible over a large area, but the slope is very steep and the underlying rock is unstable. This makes the figure vulnerable to erosion, and it needs frequent renewing. It was originally paid for by Thomas Taylor but there was no endowment for its upkeep, and maintenance had to be funded by public donations. The Bolton family, who kept the Foresters Arms in Kilburn, collected donations from patrons of the pub for its upkeep, while the Kirk family who were tenants of the land upon which the Horse was carved also had a responsibility.

After some passage of time public interest in the Horse waned, which meant that full responsibility for its maintenance had to be taken by the Kirks and the Boltons. The money raised was used to pay men to hoe and whiten the figure. However, especially during and after the First World War there was a shortage of men to do the work, while rain and melting snow caused substantial damage, and it fell into disrepair.

In 1925 the Yorkshire Evening Post started a campaign to renovate the Horse. The sum of £137 and 19 shillings was raised and it was decided to set up a team of trustees, consisting of the Revd. H. Hawkins, the vicar of Kilburn; George Bolton, a descendant of the original Bolton family; and Robert Thompson, a young man who just started making furniture in the village. They set about organising the restoration of the figure. The Post kept reporting on their progress and funds kept coming in, which paid for weeding to be carried out yearly and an occasional re-liming.

At the outbreak of the Second World War the military ordered that the Horse be hidden with camouflage netting until the war ended. It was uncovered and whitened in 1946, but an unprecedented storm almost destroyed it. That it survived is largely due to Robert Thompson and men from his workshops who were sent to weed and whiten it, "four men a full day in Spring and a full day in Summer to hoe and weed the ground at a cost of £1 per man per day, and every third year the Horse to be re-limed".

In 1955 another restoration fund was set up, and in 1973 the Kilburn White Horse Association came into being as a registered Charity. This looks after the Horse today in conjunction with local farmers. Every five years volunteers from the Association spray the limestone chippings with white paint – needing climbing ropes and harnesses to keep them in place, such is the steep gradient of the slope! Its future seems secure, although it is hoped that a more durable material will be produced to cover the Horse that will solve the problem of maintaining its whiteness.

For over 40 years the Forestry Commission had leased the escarpment and in 1955 it obtained outright ownership. A branch of the Forestry Commission - Forest Enterprise - now manages the area where the Horse is located and ensures that its setting is sensitively maintained.

Find out more

Walk to the White Horse from the National Park Visitor Centre at Sutton Bank,
http://www.northyorkmoors.org.uk/visiting/enjoy-outdoors/walking/our-walks/walking-routes/white-horse-walk

British hill-figures:
http://www.hows.org.uk/personal/hillfigs/mainwh.htm (click on 'Current Figures')

1859

The Maharajah of Mulgrave

Albert Elliot

Britain and India are worlds apart, not only in distance, but also in climate and culture. Yet despite this great gulf separating the two countries, Duleep Singh, the Maharajah of Lahore – twenty years old in 1858 – came to live at Mulgrave Castle, Lythe. How and why this came about is a fascinating story.

In 1843, at the age of five, Duleep Singh was crowned King of the Punjab and head of the Sikh nation. He was fabulously rich. However, following annexation of the Punjab by the British in 1849, he was dispossessed of all his lands, estate and possessions, including the magnificent Koh-i-Noor diamond, whose name means 'Mountain of Light'. In compensation, he was given an annual allowance of £40,000 by the British Government; a colossal sum of money in those days. At the age of eleven, the young Maharajah was exiled to Britain, going first to live with guardians in Scotland where he took up shooting and fishing, becoming a renowned marksman and angler, his Purdey shotguns his most prized possession. He was known as the 'Black Prince of Perthshire' and became a firm favourite of Queen Victoria, who befriended the good-looking young Sikh.

In 1858 Maharajah Duleep Singh moved to North Yorkshire, renting the Mulgrave Estate from the Marquis of Normanby, who was serving as British Ambassador in Florence. The Maharajah soon settled into life as an English country squire and was often seen hunting and hawking on local moors dressed in full Sikh regalia and accompanied by his extensive entourage. A contemporary account by an amazed witness verifies this extraordinary and incongruous sight near Ugthorpe:

"We were enormously aroused by a motley crew marching in line across the moors. In the centre was a fine stalwart man of some five and twenty summers, arrayed in gorgeous oriental dress – the Maharajah Duleep Singh. On either side of him were two swarthy sons of India, his Royal Falconers, with belled hawks on their shoulders, while six English gamekeepers in scarlet uniforms and a crowd of domestics filled up the picturesque tableau. It was His Royal Highness taking his sport a-hawking across these wilds. A picture, in truth, worthy of the limning of Landseer."

Queen Victoria dined with the Maharajah on at least one occasion at Mulgrave Castle, where she and her attendants were honoured guests. The Queen waxed lyrical about the Maharajah's appearance, saying he was "extremely handsome and has a pretty, graceful and dignified manner. He was beautifully dressed and covered in diamonds…" and "those eyes and those teeth are so beautiful…" Victoria, who was to become the Empress of India in 1877, remained a friend and supporter of the Maharajah and his family for the rest of his life. She became godmother to his children, even allowing Duleep's eldest son to be christened in Windsor Chapel.

While at Mulgrave Castle, the Maharajah had a new toll-road constructed between Sandsend and Whitby (the modern road follows in part the route of this). The stone-built toll-booth house can still be seen opposite Whitby Golf Course. However, the legend that he built the road to accommodate his elephants because they disliked walking along the sandy beach is not true. In fact there is no evidence of elephants at Mulgrave: it is nothing more than a myth created around the eccentric Maharajah.

Duleep Singh used to fish from a boat just offshore at Sandsend, using the age-old Chinese method of hunting with trained cormorants. These birds were tethered on a long fine lead and had a ring fitted around their necks that prevented them from swallowing any sizable fish they caught.

After leaving Mulgrave in 1862, Duleep Singh bought Elveden, a 17,000 acre estate bordering Norfolk and Suffolk, where he continued his lavish and exotic lifestyle, spending enormous sums of money on improving the house and entertaining. But in later life, as his fortune dwindled, he became dissatisfied with his lot. He visited Russia and had nefarious dealings with government conspirators and spies, hoping to enlist the Tsar's help in starting a rebellion in his homeland, with the intention of reclaiming his lands and kingdom in the Punjab. His attempts failed and sadly he would never return to his native land. He died in 1893 in Paris at the age of 55 and is buried at Elveden.

In 1999, HRH The Prince of Wales unveiled a life-size bronze statue of Duleep Singh astride his horse at Butten Island, Thetford, a town that greatly benefited from the Maharajah's patronage and generosity.

Find out more

Anita Anand and William Dalrymple,
'Kohinoor: the Story of the World's Most Infamous Diamond', by Anita Anand and William Dal (Bloomsbury 2017) - the definitive story of the Koh-i-Noor and Duleep Singh

Victoria & Albert Museum article on Duleep Singh:
http://www.vam.ac.uk/content/articles/m/maharaja-dalip-singh/

Maharajah Duleep Singh, Thetford, Norfolk

Maharajah Duleep Singh in the 1860s

Memorial to Maharajah Duleep Singh, Thetford, Norfolk

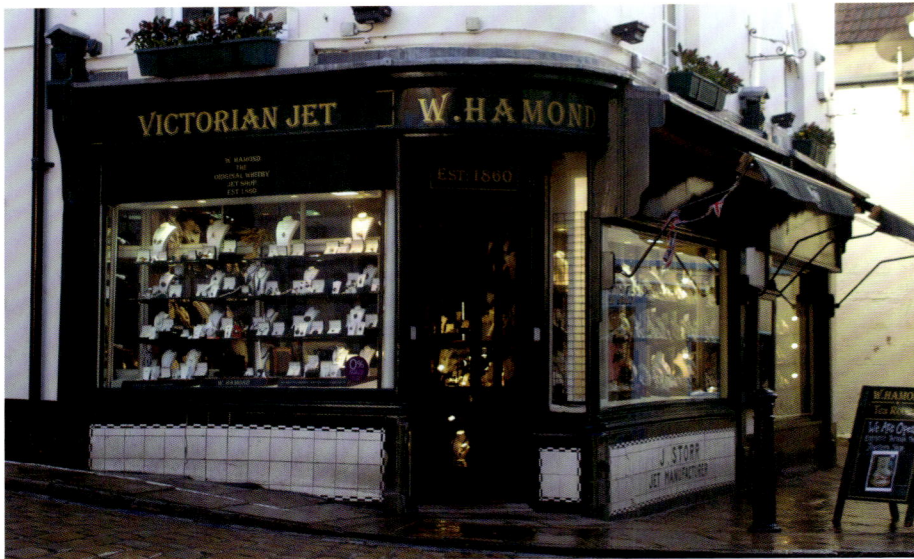
Hamond's jet workshop and shop, Whitby

Queen Victoria

Jet necklace, Whitby Museum

Monkey puzzle tree

1861

Queen Victoria and the Whitby Jet industry

Alan Staniforth

The death in 1861 of Prince Albert, Consort to Queen Victoria, not only shattered Victoria's world but brought about a surge of interest in jet jewellery.

The Queen had been married to Albert for 22 years and had come to rely on him for advice on the political questions of the day. She was also devoted to him, and his unexpected death at the early age of 42 sent her into deep mourning which was to last for over 40 years. Wearing widow's weeds for the rest of her long life, Queen Victoria complemented her dress with jet jewellery, thus bringing about a fashion in mourning apparel that lasted until after her own death in 1901.

But the history of jet goes back centuries before Victoria brought it into prominence. Writing in the first century AD, Pliny described some of the supposed uses of this dense black material found in several places around the ancient world, including in northeast Yorkshire: "the fumes of it, burnt, keep serpents at a distance and dispel hysterical affections: they detect a tendency also to epilepsy and act as a test of virginity. A decoction of this stone in wine is curative of toothache; and in combination with wax it is good for scrofula". Over time however all these uses for jet have been found wanting, and its only true value has been in the jewellery and ornament trades.

Light, dense and capable of taking a high polish, jet is a specialised form of fossil wood derived from a tree similar to the monkey puzzle (of the genus Araucaria). Whitby jet, arguably the best in the world, comes from the local Jurassic rocks where it has been preserved for up to 200 million years. It was mined extensively throughout the northern moors and coast for hundreds of years, using simple hand tools and crude mining techniques.

Unlike the local alum and ironstone industries, which had a huge impact upon the landscape, jet mining was relatively small-scale and evidence on the ground usually has to be searched for. Horizontal passages, or 'adits', were driven into the hillsides or coastal cliffs just below the strata from which alum was quarried. Waste was simply tipped out below the adit and due to its high oil content would sometimes spontaneously combust, burning the black shale to a strong red colour similar to burnt alum shale. Anyone travelling along the A172 between Stokesley and Swainby when the setting sun is striking the Cleveland escarpment will notice a clearly visible necklace of small red waste-tips, accurately marking the line of old jet workings along the contour.

Although worked in Britain since at least Bronze

Age times, it was not until the late 19th century that a jet industry evolved, based in Whitby. In its heyday, during the 1870s, there were no fewer than 200 jet workshops in the town employing over 1500 men, women and children. The jet workers' tools were simple lathes, drills, knives, files and polishing wheels. With these, a skilled worker was able to produce the most intricate of designs. But by the late 1800s jet was rapidly going out of fashion and this, coupled with imitations and cheap imports, led to the gradual collapse of the industry. A revival of interest in recent years supported by innovative new designs has led to jet once again being carved and sold in the town, albeit on a much smaller scale than yesteryear.

Walk along the shore after a high tide and you may be lucky enough to find some small pieces of Whitby jet, but beware, there is a lot of sea-coal around as well. Chemically the two are very similar so how do you tell the difference? You could try one of Pliny's tests but they are not to be recommended! It is said that if you crunch a bit of your find between your teeth and spit out black bits then you have sea-coal - while if you spit out white bits, that's your teeth and you have found Whitby jet! A much safer method is to look for the conchoidal fracture lines (like broken bottle glass) or rub a piece on dry sandstone – if the streak is black you have sea-coal, if it's brown you're in luck.

Find out more

Visit Whitby Museum, Pannett Park, Whitby YO21 1RE to view a fantastic collection of Victorian jet jewellery, a model of Whitby Abbey, and a superb chess-board with cut and polished ammonites representing the white squares
https://whitbymuseum.org.uk/

The Whitby Jet Heritage Centre, 123b Church Street, Whitby YO22 4DE,
http://www.whitbyjet.co.uk/

'Whitby Jet' by Helen & Katy Muller (2009)

1867

Alum Extraction around the North Yorkshire Moors

Michael E. Chaloner

Mineral extraction around the North Yorkshire Moors may seem always to have been an important part of life in this part of Britain, but it did have a starting point.

If we go back to 16th century Britain, one of the country's major industries was producing woollen clothing – yet there was a major problem in the industry, which was fixing colours in these woollen goods. In the past various substances had been used as a mordant – including urine – but by then it had been discovered that the most effective means was using alum. Alum is a mix of the crystalline salts of Potassium aluminium sulphate and Ammonium aluminium sulphate. However, at that time the alum which was so essential had to be imported from Italy, which had a monopoly on European production. Italy was then effectively governed by the Roman Catholic Church, but these were Elizabethan times, when Protestant Britain was seen as an important enemy of the Catholic Church, and so the release or withholding of alum could be used as a financial and political weapon.

It was clear that it was important to find a source of alum in Britain to overcome this problem. In the late 16th century Sir Thomas Chaloner the Younger (1559-1615) was on a tour of Europe when he visited the alum extraction plants in Italy. He appreciated that the local plant life was similar to that on his Guisborough estate, and thought it worth trying to extract alum from the shale deposits found in the Guisborough area. He called in the support of his cousin Thomas Chaloner of Lambay (a small island near Dublin) who in 1595 managed - with the help of alum workers from Germany or Italy - to develop a process that brilliantly overcame the difficulty of extraction. The process involved roasting the shale for between 6 and 9 months to convert the sulphides and oxides into sulphates.

The level of heating and length of roasting was critical and a considerable achievement to get correct at a time when the science of chemistry and therefore the understanding of these processes did not yet exist.

Thomas Chaloner of Lambay was awarded a Government pension (40 marks per year) for the achievement, but when he was 70 years old he had to walk over 250 miles to London from Guisborough in order to persuade the government of Charles 1st to continue to pay the pension: the Chaloner family had trouble getting money from the King after he had forced them to lease the workings to him.

Essential elements of the alum production process had to be imported from elsewhere in Britain, especially the coal needed as fuel, urea (from urine) as a source of ammonium to precipitate formation of the alum crystals, and dried seaweed as a source of potassium. The coal was generally brought in from Newcastle and the urine from London, since production of this fluid from local cities such as Newcastle and Hull proved insufficient. In a useful way of helping to alleviate the challenge of sewage disposal in the increasingly populated capital city, Londoners were paid for their urine, which was transported north in wooden tubs – with the empty tubs sent back to London containing Yorkshire butter! As these resources were mostly transported by sea, the larger scale extraction works tended to move closer to the coast over the years. In fact, the alum industry was a significant factor in stimulating the rapid expansion of Whitby's shipping industry during the 17th century.

Eventually there were at least 17 alum works. In the heyday of the industry workings took place along the North Yorkshire coast at Loftus, Sandsend and Ravenscar. The last working mine was at Sandsend, where extraction of alum came to an end in 1871. By this time simpler methods of manufacture had been developed and the woollen industry was no longer so important, so there was less demand for the mineral. However, the chemically related Potassium sulphate is still extracted from under the northeastern edge of the North Yorkshire Moors, at Boulby. Meanwhile, the Chaloner family moved on to ironstone extraction from their Guisborough lands to support the Teesside steel plants, with the Chaloner pit operating until 1939.

Many places along the Yorkshire coast still show evidence of the alum industry, including huge piles of shale discarded after the extraction process. Former workings can be seen at the Loftus Alum Quarries and the Peak Alum works at Ravenscar, near Robin Hood's Bay, which can be visited on foot from the National Trust's Coastal Centre at Ravenscar.

Find out more

The Cleveland Way footpath offers a beautiful route for visiting ten sites of the old workings
http://www.teeswildlife.org/what-we-do/past-projects/alum-alchemy-and-ammonites/places-to-visit/cleveland-way-alum-sites-guide/

'Thomas Challoner and his Astonishing Alum Industry' by Adam Hart-Davis (1995),
at http://www.exnet.com/1995/12/18/science/science.html

Headland quarried for alum shale, Sandsend

Robert Thompson's trademark carved mouse

Robert Thompson in his workshop

Work-bench, Mouseman Museum, Kilburn

Oak stacked and seasoning, Kilburn

1876

Robert 'Mouseman' Thompson, Master Craftsman from Kilburn

Adrian Leaman

The best view in England is often said to be from Sutton Bank in North Yorkshire. Directly west from your viewpoint is Thirsk, then Wharfedale and the Pennines beyond. The escarpment below your feet has oak woods; to the south, about a mile away, is the village of Kilburn. Behind you, out of sight and to the southeast, is Ampleforth Abbey. On a clear day, you may be just able to make out York Minster.

If you stretch your imagination, Oxgodby, the fictional village in J. L. Carr's elegiac classic 'A Month in the Country', is also in your line of view. In the novel, the Oxgodby chapel-goers' Sunday summer outing takes them to the village of Kilburn. In the main street a joiner calls to someone he recognises, as the trippers' charabanc passes his front yard. If you suspend disbelief a little more, that joiner could be Robert Thompson. It is 1920. He's thinking about the new furniture he is making for Ampleforth Abbey. He is toying with the notion of adding a carved mouse ...

Towards the end of his life in 1949, Robert Thompson (1876-1955) wrote: *"The origin of the mouse as my mark was almost in the way of being an accident. I and another carver were carving a huge cornice for a screen and he happened to say something about being as poor as a church mouse. I said 'I'll carve a mouse here' and did so, then it struck me what a lovely trade-mark. This is about 30 years ago".* That mouse has gone on to make Robert Thompson the most famous of many fine carvers of English oak. And churches proved to be enduring clients for his company, and remain so for his descendants today. The mouse works as a trade-mark, one of the earliest to survive into the modern day, but also helps authenticate the work of Robert Thompson's Craftsmen Ltd, and does service as a benchmark for quality. The mouse also serves perfectly for everyone wishing to seek out and spot examples of Mouseman furniture and carved works.

Carving a mouse in English oak is a demanding test of skill and patience. Oak combines hardness, weight, suppleness and toughness, but none of these to extremes, making it ideal for furniture and other uses requiring strength and durability, most notably medieval shipbuilding and construction. But a mature oak takes 120 years to grow, and is thus expensive. Oak is heavy and cumbersome, so ready availability of local supplies is important, especially when access is a challenge. Oak needs to dry out slowly, hence the large rectangular piles that may be seen in Thompson's Kilburn yard.

To hand-work oak, you need tools for paring and gouging, and specialist tools such as veiners, with a U-shaped edge. You need mallets and chisels and, famously, in the case of Robert Thompson, the adze, a smaller version of a mattock, with a sharp cutting edge that shaves the wood's surface, rather than slicing or cutting it. Evidence of flint adze work has been found at Starr Carr, near Scarborough, Britain's earliest known house, at 11,000 years old - so such techniques have been around a long time in Yorkshire. Use of the adze produces the characteristic hexagonal ripple effect in items such as refectory tables. Hand-working oak takes considerable time and effort, but the end result is unparalleled in domestic furniture such as sideboards and dressers, panelling, carvings, ecclesiastical fixtures and fittings such as pulpits, pews and prayer desks. There are even Mouseman clocks and barometers, and smaller items such as ashtrays, fruit-bowls and book-ends.

Today, people wishing to go on a 'mouse-hunt' can do no better than starting at St Mary's, the charming village church of Kilburn, tucked away behind Thompson's yard. Other examples include the church of St Mary the Virgin in Norton-on-Tees, where the lych-gate and pews have 'mice'. Visitors may see some of the latter-day craftsmen at work at the Mouseman Visitor Centre in Kilburn, where there is also a splendid tea-room, complete with Mouseman dining chairs and bespoke café tables. The furniture range is displayed in Robert Thompson's original cottage, now serving as a showroom. But be warned: it's tempting!

(With thanks to Ian Thompson Cartwright for additional material.)

Find out more

Mouseman Visitor Centre, Kilburn YO61 4AH,
http://www.robertthompsons.co.uk

Motoring trails are detailed in 'Mouseman:
The Legacy of Robert Thompson of Kilburn', by Patricia Lennon and David Joy (2008)

The delightful 1948 silent film 'Craftsman of Kilburn' can be viewed at
http://www.yorkshirefilmarchive.com/film/craftsman-kilburn

1880

Frank Elgee, Man of the Moors

Albert Elliot

High on the moors near Ralph Cross on Blakey Ridge, virtually at the centre of the National Park, is a memorial stone unveiled on 21st October 1953 by Harriet Elgee, the widow of Frank Elgee. The inscription reads simply 'Frank Elgee 1880–1944. Naturalist. Archaeologist'. All the leading natural history and archaeological societies donated funds towards the memorial in recognition of the great contribution Elgee made to the area. This simple stone commemorates the life of an exceptional man who was passionate about the North Yorkshire Moors.

Frank Elgee was born at North Ormesby, Middlesbrough. In 1888 he caught scarlet fever and throughout his childhood suffered from illnesses which eventually wrecked his health, curtailing his formal academic studies at the age of 14. In 1895 he became an office boy. However, the confined environment and long working hours exhausted him and his health broke down completely, leaving him seriously ill. He underwent a major chest operation at North Riding Infirmary, in Middlesbrough, and although after a few weeks the sickly teenager was allowed home, the prognosis was bleak and his future uncertain.

His parents took their chronically ill son for recuperation to Ingleby Greenhow, at the foot of Urra Moor. Confined to a bath-chair while convalescing there, he read voraciously and studied. Although in poor physical health, his enquiring mind was insatiable for knowledge and he was, even then, determined to discover the origin and evolution of the northeastern moorlands, whose blue escarpments were tantalisingly in view from his sick-bed. Elgee went on to spend a lifetime studying the natural history and archaeology of his chosen area. As he grew stronger, he paid frequent visits on foot to his beloved moorlands to carry out studies and investigations first-hand, eventually becoming intimate with the whole of the northeastern moorland. He was meticulous and kept careful records of his travels and investigations.

In 1904 he was appointed Assistant Curator (and later Curator) to the Dorman Museum, Middlesbrough, a post he held from 1904 to 1938. This enabled him to do carefully planned work and gave him the wherewithal to travel further afield, staying overnight at farmhouses on remote moors. The whole of the area, some 400 square miles, was now open to his eager scrutiny. He began keeping diaries. In 1907 he started writing his first major book, 'The Moorlands of North-Eastern Yorkshire; Their Natural History & Origin'. Published in 1912, the book was a pioneering original work and the first ever in-depth regional survey to be published in Britain.

In 1911 Elgee met Harriet Wragg, who was to become his wife and assistant, in Danby. From 1920 to 1931 Frank and Harriet lived in the moorland village of Commondale, from where he travelled to work along the Esk Valley railway line and wrote his second book, 'Early Man in North-east Yorkshire' (published 1930). His profound respect for and pride in the area shine out from lines taken from the preface:

"This work surveys the archaeology of North-east Yorkshire, one of the greatest prehistoric regions of England... With this region I have been on intimate terms for a lifetime so that it is not too much to say that my love for it has been a powerful motive in the creation of this work. Otherwise I question whether I should have been able to summon up enough patience, endurance and courage to study so much arid archaeology, or to examine dusty antiquities in so many museums. These dry labours, however, were essential to a right understanding of the vast collection preserved on the wide-open spaces of the moors, the North-east's greatest glory. Here amongst the bracken and ling and with the companionship of wind, sun and rain, archaeology became a real joy and the life of the past a real presence."

His third major book, written in conjunction with Harriet, was 'The Archaeology of Yorkshire' (published 1933). Also in 1933, the local and national importance of his work was recognised by conferment of the degree of Doctor of Philosophy by Leeds University. Due to ill-health, he resigned as curator of the Dorman Museum in the same year and was succeeded by Harriet, who held the post until 1938. Elgee remained forever grateful for what his beloved moors had given him and wrote the simple but moving words: 'The moors have satisfied my reason, captivated my imagination, and elevated my heart'. His health continued to deteriorate, so the couple moved to Alton, in Hampshire, in search of a milder climate. He died in 1944 aged 66 and is buried at Alton.

Find out more

Frank Elgee's 1912 book 'The Moorlands of North-eastern Yorkshire' is online at
https://archive.org/details/moorlandsofnorth00elge

Memorial Stone on Blakey Ridge at the junction with the road to Rosedale Abbey, grid ref. NZ 676019

The Dorman Museum, Linthorpe Road, Middlesbrough TS5 6LA,
http://www.dormanmuseum.co.uk/

Frank Elgee

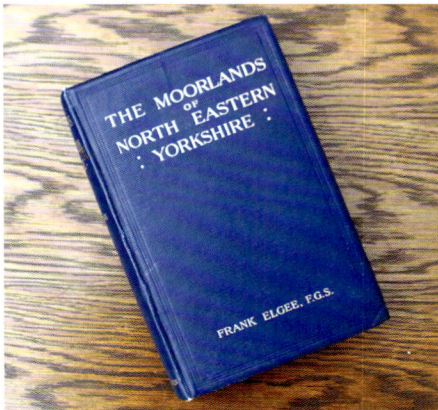

Elgee's first book

Memorial stone to Frank Elgee on the moors, near Ralph Cross

The view down Rosedale from close to Elgee's memorial stone

On the 18th January 1881 the Brig
"VISITER"
ran ashore in Robin Hood's Bay. No local boat could be launched on account
of the violence of the storm, so the Whitby lifeboat was brought overland
past this point — a distance of 6 miles — through snowdrifts 7 feet deep on a
road rising to 500 feet, with 200 men clearing the way ahead and with
18 horses heaving at the tow lines, whilst men worked uphill towards them
from the Bay. The lifeboat was launched two hours after leaving Whitby
and at the second attempt, the crew of the Visiter were saved.
So that future generations may remember the bravery of Coxswain Henry
Freeman, and the lifeboatmen, and the dogged determination of the people
of Whitby, Hawsker and Robin Hood's Bay, who overcame such difficulties;
this memorial was erected in 1981.

Scene of the Visiter disaster below the distant Ravenscar headland, Robin Hood's Bay; Inset, commemorative plaque

1881

The 'Visiter' Sea Rescue, Robin Hood's Bay

Jane Ellis

A plaque at the top of The Bank at Robin Hood's Bay tells the story of a heroic rescue which took place in 1881, when a collier brig the 'Visiter' (though in contemporary news reports the spelling was 'Visitor') foundered in a violent storm whilst carrying a cargo of coal from Newcastle to London. The vessel, registered at Whitby and locally owned, was by then elderly, having been built in 1823 at Sunderland. She had sailed as far south as Flamborough by the afternoon of Tuesday 16th January when a south-easterly gale prevented further progress, tearing the sails to shreds and driving her back up the coast past the cliffs of what is now known as Ravenscar, though in those days was called Peak.

In the middle of the night she was taking on so much water that the master set down the anchor, hoping to ride out the storm. The wind veered north-easterly bringing snow and hail, and sea conditions were atrocious. Waves were breaking over the deck and the crew of six attempted to save themselves by taking to the ship's boat, but dared not leave the comparative shelter of what was by now a wreck for fear of being driven onto the unforgiving rocks beneath the cliff. Dodd, the apprentice, was wet through and frozen, having jumped into the sea and swum to the boat roped to a buoy, to be pulled into

it by the other crewmen, and here they spent the bitter winter night in conditions beyond imagination.

It was only when the brig's quarter-board was found on the beach at Robin Hood's Bay next morning that anyone realised there was a shipwreck. The six men in their life-threatening predicament could just be seen from land two miles to the south of the village. The Bay lifeboat was old and the local fishermen who were its crew regarded it as unseaworthy for launching into such mountainous seas, a decision confirmed by the coastguards who inspected it. In desperation the Vicar, Revd. Jermyn Cooper, sent a telegram to Whitby: "Vessel sunk, crew in open boat riding by the wreck, send Whitby lifeboat if practicable".

Rowing the lifeboat, the Robert Whitworth, around the coast from Whitby in such conditions was out of the question, but the Whitby branch secretary of the Royal Naval Lifeboat Institution (RNLI), Captain Gibson, along with first and second coxswains Henry Freeman and John Storr, took the momentous decision to haul the boat the six miles to Bay overland. This would have been difficult enough in fair weather, with the road climbing steadily to an elevation of over 500 feet followed by a steep drop down to sea-level, but the whole country was at the time suffering from severe winter weather, with hard frost and blizzards which had blocked many roads. With the combined efforts of 18 horses and over 200 volunteers who turned out in the dreadful weather to cut a way through snowdrifts up to 8 feet deep, Bay remarkably was reached in just over three hours. The way down the steep and twisting Bay Bank was difficult and dangerous, with the helpers controlling the descent by means of ropes attached to the boat's carriage, and as she rounded the double-bend by The Laurel Inn there was barely an inch and a half to spare.

The lifeboat was launched successfully, albeit with some difficulty, however before the wreck could be reached, six of the oars were snapped by one tremendous wave and the lifeboat had to return to shore. Some of the oarsmen were by now too exhausted to row, but following an appeal made by Henry Freeman for volunteers, the boat's second attempt - now with a larger crew and using the oars from the old Bay lifeboat - was successful.

The shipwrecked men were suffering badly from exposure, two of them by now delirious, but with the sea still raging all were somehow manhandled into the lifeboat. The dramatic rescue was completed by mid-afternoon on the Wednesday and thankfully all six unfortunate men escaped with their lives.

Several days later, the storm having abated, the lifeboat crew walked the six miles from Whitby back to Robin Hood's Bay and rowed the Robert Whitworth around the coast back to Whitby harbour. Later that year, the RNLI provided Bay with a 32-foot self-righting lifeboat, the Ephraim & Hannah Fox, together with a brick lifeboat-house in the Dock which stands to this day. It bears a record of all the rescues carried out by the station until closure in 1931.

Find out more

Scarborough Maritime Heritage Centre, 45 Eastborough, Scarborough YO11 1NH,
http://www.scarboroughsmaritimeheritage.org.uk/

Robin Hood's Bay,
https://www.robin-hoods-bay.co.uk/

1890

Bram Stoker, Dracula and Whitby

Fleur Speakman

Bram Stoker's novel 'Dracula', published in 1897, influenced numerous novels, films, plays, and at least one ballet. Gothic horror novels with a vampire theme had already proved popular in England in the 18th century, though the blood-sucking vampire had apparently much earlier antecedents in folklore, particularly in Ancient Egypt, Asia, and many other regions throughout the world.

Born in Ireland in 1847, Stoker, who suffered poor health in childhood, was entertained by his mother with supernatural stories from her native Sligo, in which the vampire is likely to have featured. After studying mathematics at Trinity College, Dublin, Stoker became a writer initially of non-fiction, then added theatre reviews and two novels with an Irish setting to his literary output. One of his reviews impressed the celebrated actor-manager Henry Irving, and their friendship resulted in Stoker becoming Irving's business manager for 27 years. In 1890, after an exhausting but unsuccessful tour of Scotland by the theatrical troupe, Irving suggested a month's holiday for Stoker on the coast, recommending Whitby. Stoker was enchanted with the stunning setting of the town - viewed from the West Cliff where he had lodgings in the Royal Crescent.

Inspired by Whitby, Stoker started work on a new novel, initially set in Austria, with Count Wampyr as its central character. The Gothic literary tradition of eerie castles, huge forests, unknown customs and a foreign language helped to increase the tension of the supernatural terrors unleashed on the reader. But Stoker realised that Whitby itself could provide the perfect backdrop for much of the drama, with its amazing windswept headland, ancient ruined Benedictine Abbey on the site of the 7th century original, and spectacular 199 steps up to St Mary's Church on the East Cliff. Haunted by bats in the evenings and with some atmospheric worn gravestones, there was even a convenient local legend of a white lady flitting by one of the Abbey windows. A number of other tales were related to him by local folk and by seafarers who put in to port.

Researching in Whitby library, Stoker found details of the exploits of a 15th century prince called Vlad Tepes, who impaled his enemies on wooden stakes. Vlad was known as 'dracula', meaning 'devil' in the Wallachian language, but it was also an admiring surname to denote conspicuous courage, cruel actions and cunning. These qualities were later embodied in Stoker's shape-shifting Count Dracula.

Stoker's first version of his story was a play, a vehicle for Irving, called 'The Undead'. But Irving remained unimpressed. Six years later, after much additional research, Stoker produced his novel 'Dracula', though he never actually visited Transylvania, where much of rest of the action is set. The finished story is in the form of letters, diaries, newspaper cuttings and even a ship's log – all giving different viewpoints. A mysterious boat, the Demeter, arrives in Whitby in the fog, without passengers or crew but with a dead sailor lashed to the mast, while a strange black dog jumps ashore from the wreck and vanishes - actually a disguised Count Dracula. In reality, an actual Russian ship had run aground on Tate Sands below the East Cliff with a cargo of silver sand some years before; Stoker's ship too had some silver sand as cargo.

A key character in the novel, Doctor Van Helsing, scientist and psychic doctor, uses a blend of modern science, coupled with religious artefacts and bulbs of garlic, to defeat Dracula. Stoker is also careful to paint a repellent picture of the sleeping vampire: "it seemed as if the awful creature were simply gorged with blood; he lay like a filthy leech, exhausted with his repletion". Once caught, Dracula suffers death by the traditional stake through the heart and decapitation.

Particularly fascinating is the novel's use of contemporary technology to further the plot, such as a phonograph with wax cylinders and a portable typewriter able to produce carbon copies. It is noteworthy that Mina, quite a courageous period heroine, skilled in shorthand and typing and in making transcriptions from the phonograph, helps to clarify key happenings. But her role is clearly subservient to the male patriarchy - the men of action. The novel's sub-text is of particular interest. Dracula's hold over the two key female characters, aided by his "deep-burning eyes", suggest his powerful mesmeric sexual attraction. His repeated blood-sucking forays, by biting his victims' necks, doom them to become vampires, preying forever on others and totally under his control.

Find out more

Ruins of Whitby Abbey, Whitby YO22 4LT

Whitby Ghost Walks,
https://www.whitbywalks.com/

Whitby Goth weekends, twice per year,
http://www.whitbygothweekend.co.uk/

'Dracula' by Bram Stoker (1897, many subsequent editions)

Abraham "Bram" Stoker

Bram Stoker's stay at 6 Royal Crescent, Whitby

Whitby Harbour from East Cliff

Whitby Abbey at dusk, East Cliff, Whitby

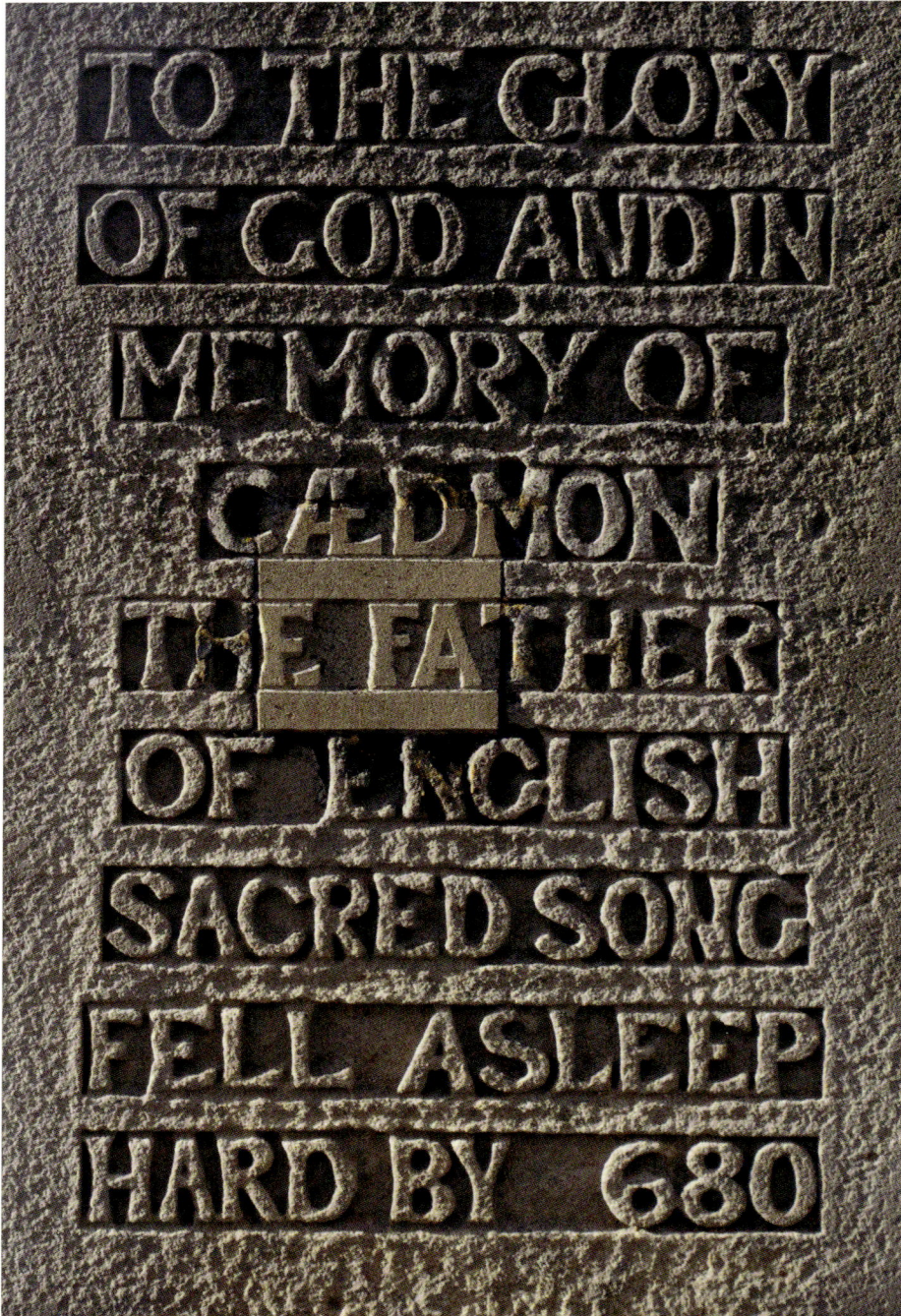

Caedmon Cross inscription

Caedmon Cross, St Mary's churchyard, Whitby

TO THE GLORY
OF GOD AND IN
MEMORY OF
CÆDMON
THE FATHER
OF ENGLISH
SACRED SONG
FELL ASLEEP
HARD BY 680

1898

Caedmon's Cross in Whitby

Sharon Artley

'Cademon' or 'Cadmon'? The 7th century Anglo Saxon poet is known locally as 'Cademon', but further afield is widely referred to as 'Cadmon'.

Caedmon was a herdsman who lived at Whitby Abbey during the time of Abbess Hilda. Whitby was then a double monastery, where communities of nuns and monks lived alongside each other. It was a great centre of learning and in AD 664 hosted the Synod of Whitby, where it was finally decided when the date of Easter should be celebrated.

In addition to the nuns and monks there was a lay community, and Caedmon, a lay brother, looked after the cattle within the grounds. His story can be found in Bede's 'Ecclesiastical History of the English People'. Bede describes Caedmon as a man without learning, and how when attending banquets after which there was entertainment when those present were expected to sing: "if he saw the harp come towards him, he would rise up from table and go out and return home".

On one such occasion, having left the banqueting hall, he fell asleep in the stable and had a dream in which he was asked to sing. Caedmon claimed he could not, but the person in his dream persisted with the request. When Caedmon asked what he should sing about, he was told to "sing the beginning of creation". He did so in English and

astonishingly, when he awoke he remembered his song.

He told the reeve (a monastery official) who promptly took him to Abbess Hilda. She believed that Caedmon had been given a divine gift, and encouraged him to take monastic vows and be received fully into the monastery. Bede describes that through his learning, Caedmon was able to turn many stories from scripture into song. Unfortunately, only his initial 'Song of Creation' has survived.

In October 1897, following an initiative by the Revd. Canon Rawnsley of Keswick (a founder of the National Trust), a public meeting was held and a committee appointed to create a memorial cross which would celebrate Caedmon's contribution to sacred literature in English. Subscriptions were invited and approximately £250 was raised, including a donation from America.

On 21st September 1898, a ceremonial unveiling of the new cross took place. There was a large gathering including the Poet Laureate Alfred Austin, bishops and other clergy and local dignitaries. Canon Rawnsley's address made mention of the word of God being borne through the poems of Caedmon to "the hearts of wild Northumbrians and the homes of the dwellers upon Danby Moor".

The cross is almost 20 feet high and made of hard sandstone. The cross-head on the east side is carved with the Lamb of God (Agnus Dei) together with the four Evangelists and their symbols. This side faces the Abbey and has four panels down its length. At the top is Christ in the act of blessing, King David playing a harp, the Abbess Hilda, and finally Caedmon in the stable being inspired to sing his hymn. At the base is a plaque which reads: "To the glory of God, and in memory of Caedmon, Father of English sacred song. Fell asleep hard by AD 680."

On the west face, the cross-head has a raised boss in Celtic knotwork and a dove symbolising the Holy Spirit. Down this face is a double vine symbolising Christ.

In the loops of the vine are four scholars working in the Abbey at the time of Caedmon. Underneath this is carved the first nine lines of Caedmon's 'Hymn of Creation' in English. The two narrower sides of the cross depict a design of a rose, birds, animals and an apple tree symbolising Eden. In pairs of letters in two scripts – Celtic uncial on one face and runes on the opposite – is Caedmon's Hymn of Creation in his native Anglo-Saxon. (Letters written in an uncial script are more curved, whereas runes are made up of straight and angled lines.)

A sealed glass bottle to date the event was counter-sunk into the base of the shaft. It contained silver coins of the year, a photograph of Queen Victoria, the names of subscribers and the sermon given by Canon Rawnsley.

Many years later, Mrs Mabel McMillan, a Whitby councillor, expressed concern that there was no memorial to the Father of English verse outside his home town. This was rectified in 1966, when in the 900th anniversary year of the foundation of Westminster Abbey a memorial to Caedmon "who first among the English made verses" was unveiled in Poets' Corner.

Find out more

St Mary's Church, Whitby YO22 4JR

Caedmon's 'Hymn of Creation' in Old English and in translation,
http://www.thehypertexts.com/C%C3%A6dmon's%20Hymn%20Translation.htm

1904

Henry Freeman, Man of The Sea

Ray and Sheila Clarke

Henry Freeman was born on 29th April 1835 in Bridlington to William, a brick-maker, and Margaret; he was one of eleven children. When he was five the family moved to Flamborough, a place of crashing seas and stormy winters, which must have had a profound effect upon the young Henry. Along with his brothers Henry was instructed by their father in the brick-making trade, something most of the boys took up. Henry initially followed his older brother William into farming, but returned to brick-making when he moved to Whitby in 1855.

History records that in April 1858 Henry chanced to look out to sea and espied a traditional wooden fishing-boat – a coble - upset by waves between Upgang and Sandsend; was this the moment when he decided to pursue a life at sea? For three years he worked on a variety of coastal vessels before returning to Whitby to become a fisherman. Then an event took place which shaped the rest of his working life.

The evening of 8th February 1861 saw a coastal assault by exceptionally strong north-easterly winds, whipping up the seas into a wrath of anger. Over a 24-hour period more than 200 vessels are believed to have foundered. Although he had not worked on lifeboats before, Henry was involved in five rescue attempts, but during the fifth rescue the lifeboat was overturned by two enormous incoming waves colliding with a third on its rebound; the resultant explosion of force upended the lifeboat, causing its crew to be tipped into the sea. All drowned bar Henry; they were wearing the then standard lifebelts that fitted low on the body whilst he was wearing a newer one that fitted over his shoulders. Henry was 26 at the time of the tragedy and went on to serve as a lifeboatman for 40 years, of which 22 were spent as coxswain.

Away from his lifeboat duties Henry was a fisherman. From July to October he would typically set nets from a coble, catching herring, and use long-lines with multiple hooks to catch mackerel. In the 1870 register of Whitby Fishing boats, we see Henry achieving his goal of taking on a boat as master.

Henry Freeman was not a man without controversy, taking it upon himself to launch when others might not have done. In 1881 he was involved in the dramatic rescue of the crew of the 'Visiter'. He became an ambassador and spokesman for the Whitby fishing community, for instance campaigning for coloured warning lights on fishing trawlers. He was an energetic promoter of the RNLI (Royal Naval Lifeboat Institution - founded in 1824). He travelled to London, Leeds, Halifax and Huddersfield, including occasions when he and a crew would parade through the city aboard a lifeboat; this brought much needed funds to the RNLI. Throughout his time with the lifeboat service Freeman gained much acclaim for his daring exploits - and his reputation was to stand him in good stead when he came up against the law.

In April 1883 Henry and his three partners were charged with the theft of fishing lines set at sea, an offence of such seriousness that imprisonment was not uncommon; for Henry his good name and reputation were also at stake. He was found guilty and given a hefty fine of £15, his previous good character having saved him from a more severe sentence. He retained his position as coxswain until his retirement in 1898, a year of great upheaval for him since he also lost his wife.

In 1861 Henry had married Elizabeth, daughter of Thomas Busfield, a jet ornament maker. The marriage lasted 36 years, although sadly there were no children. Elizabeth succumbed to a liver disease in 1898 and was cared for by her sister Emma, a widow of some 30 years. In 1901 Henry and Emma married and once again Henry ran foul of the law: since 1835 it had been illegal for a man to marry his wife's sister, whatever the circumstances. Both participants must have been aware of what they were undertaking, for they took the unusual step of marrying in Hartlepool. The Act would not be changed until 1907.

On 13th December 1904 Henry Freeman died. His funeral was a well-attended affair at which his coffin was borne by his fellow lifeboatmen, with a large crowd following. During his time with the lifeboats he had saved over 300 lives.

Find out more

The Whitby Museum, Pannett Park, Whitby YO21 1RE, https://whitbymuseum.org.uk/ ,
has a large, imposing portrait of Henry Freeman, donated by the Graham family.

The Sutcliffe Gallery, 1 Flowergate, Whitby YO21 3BA, http://www.sutcliffe-gallery.co.uk/
has photos of Henry Freeman Scarborough Maritime Heritage Centre, 45 Eastborough, Scarborough YO11 1NH,
http://www.scarboroughsmaritimeheritage.org.uk/

Whitby Lifeboat Museum, Pier Road, Whitby YO21 3PU -
https://rnli.org/find-my-nearest/museums/whitby-lifeboat-museum

'Storm Warrior: Turbulent life of Henry Freeman' by Ian Minter and Ray Shill (1991)

Henry Freeman with the new design of cork life-jacket that saved his life

SS Rohilla, by Bill Wedgewood

Dragging the No. 2 lifeboat, by John Freeman

The Rohilla foundered at Saltwick Nab, a mile from Whitby Harbour

Within 50 yards of the wreck, by Diana Moore

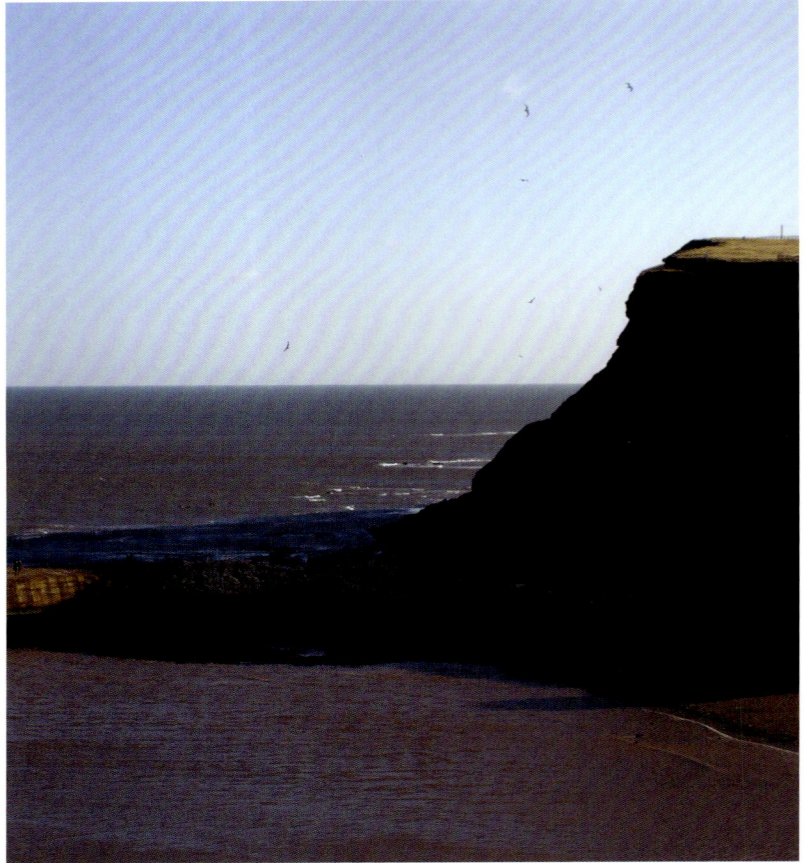

Trying to launch the William Riley, by Tricia Shaw

1914

The Wreck of the Rohilla

Sue Morton

In World War I the SS Rohilla was commissioned as a hospital ship, working from her home port of Leith in Scotland. Rohilla's Captain, David Neilson, was a very experienced sailor and, with his crew, had been with the ship since the day of her commissioning - but he had never sailed the North Sea before.

On Wednesday 28th October 1914, HMHS Rohilla was ordered to Dunkirk to evacuate wounded soldiers from the British Expeditionary Force. There were 229 people on board including doctors, five nurses and orderlies. At dusk, as the ship headed down the east coast of England, a fierce storm arose and soon the Rohilla was battling against gale-force winds and heavy seas. Due to wartime blackout regulations, no coastal, harbour, maritime or ships' lights were allowed. So in total darkness, in high winds and raging seas, the captain could only navigate by dead-reckoning.

By 3.30am on 30th October, the Rohilla neared Whitby. Although Captain Neilson had calculated that the ship was 7 miles from shore, she was actually less than half a mile from Whitby's perilous rocky coastline. The Coastguard at Hawsker spotted her and urgently signalled warnings in Morse code, sounded the foghorn, sent out maroons (warning rockets) and called out the Whitby lifeboats. At 4.00am, at the change of watch, a crew member sent to take soundings told the Captain he had noticed Morse signals in the sky above the ship. But before a signalman could arrive on the bridge, the ship struck the Whitby Rock at full power and then rode onto the notorious East Scaur, just 450 yards from the shore. The bows and mid-section balanced on the rock but the stern was torn away, drowning 60 poor souls trapped there.

Rockets carrying lines were fired towards the ship in an attempt to evacuate the remaining passengers, but the wet, heavy lines failed to reach the ship, while huge seas breaking over the piers prevented Whitby's two rowing lifeboats from launching through the harbour mouth. As dawn broke, the lifeboat coxswain ordered the No 2 lifeboat to be carried over the East Pier's 8-foot-high sea wall. It was dragged for a mile along the scaur towards the wreck site by 100 men and women. The lifeboat rescued 33 people in two hazardous journeys before becoming so damaged that she was abandoned. All but one of Rohilla's lifeboats had been wrecked. She launched the remaining boat only for it to be smashed against the ship's side.

Meanwhile the Upgang rowing lifeboat, the William Riley, was transported across the cliffs, through Whitby and over to Saltwick by 6 horses and 100 men. With winches and ropes the boat was lowered 200 feet down the cliffs to the shore, only to discover that the onshore gale and lashing seas prevented launching. People from miles around travelled to the cliff-tops to offer help and watch in horror.

The Teesmouth lifeboat was holed leaving harbour and had to return to base. The Scarborough rowing lifeboat, towed by a steam trawler, arrived at 6 pm and stood by all night but, being unable to reach the ship by early morning, returned to port with her exhausted crew. On the stricken ship the passengers, unaware of the unflagging and brave attempts being made to rescue them, feared they had been abandoned.

Throughout that terrible Saturday, the Whitby lifeboats, aided by the steam trawler Mayfly from Hartlepool, continued to make rescue attempts. At last the William Riley was launched from the shore. The crew rowed her within 50 yards of the wreck and in desperation people jumped overboard, trying to reach the lifeboat or the shore. Some were dashed onto the rocks and drowned.

Local men and women gathered on the shore or waded into the raging seas, repeatedly risking their own lives, to save others.

Tynemouth's steam lifeboat Henry Vernon arrived in Whitby harbour at 1 am on Sunday 1st November and took on barrels of oil. The boat got within 200 yards of the wreck and poured the oil on the waters, flattening the heavy seas. In 15 minutes, the 50 remaining men were taken on board, including Captain Neilson, carrying the ship's cat. The Henry Vernon sailed into Whitby to cheering crowds and ringing church bells.

After enduring over 50 hours of exposure, in gale-force winds and terrifying seas, the lifeboats and their brave crews had saved 146 lives. The various seamen, coastguards and local people from several communities had risked their own lives to effect a brave and remarkable rescue.

Find out more

Whitby Lifeboat Museum, Pier Road, Whitby YO21 3PU,
https://rnli.org/find-my-nearest/museums/whitby-lifeboat-museum

Remarkable contemporary newsreel of the rescue,
https://www.youtube.com/watch?v=qUzIw_RS0qk

Website on all things to do with HMHS Rohilla,
http://www.eskside.co.uk/ss_rohilla/index.htm

1919

The Forestry Commission and the North Yorkshire Moors

Brian Walker

By the 17th century the coverage of woods and forests in Britain had declined to less than 12% of the land area. In 1900 this had fallen further to only 5%, and the nation's industries relied on imported timber. But with the outbreak of World War 1 timber imports could no longer be relied on, and the nation's timber stocks were depleted even further. Huge quantities of timber were not only needed for pit-props in the coal mines but also in the trenches on the front line in France and Belgium, and as sleepers for the hundreds of miles of railway lines transporting troops, ammunition and other supplies.

By the end of the War tree cover had fallen to about 3% of Britain's land area. The Government recognized that there was a strategic, critical shortage of trees and in 1916 had already appointed the Acland Committee to look into developing new woods and forests. In 1918 the Committee presented its findings and proposals for a Forestry Commission were accepted. The Forestry Act received Royal Assent on 1st September 1919.

The Forestry Commission was tasked with protecting remaining timber supplies and creating new stocks through the acquisition of land for planting new forests, restrictions on felling trees, and incentives to private landowners to create more woods and forests. It was also charged with carrying out research into everything associated with trees, woods and forests: from the soils they grew on to the bugs that eat them and the birds and other animals that live in them.

This was a mammoth task, the like of which had not been seen anywhere in the world before. There was little money to buy and manage the new forests. However, marginal agricultural land was cheap following the War and large acquisitions were made, mainly across eastern and northern England and in Scotland and Wales. Some traditional Crown forests such as the New Forest and the Forest of Dean were also transferred to Forestry Commission management.

On the North Yorkshire Moors the first acquisitions were at Low and High Dalby. This former farming and rabbit warrening land was purchased from the Duchy of Lancaster Estate in 1919 and tree-planting began in 1921. Some of the first trees planted are still growing alongside Thornton Beck, south of Low Dalby. Over the following decades the Forestry Commission added to its holdings on and around the Moors: it now cares for over 21,000 hectares in the area.

1919 to 1960 saw profound landscape changes, with forest returning to places that had been deforested in prehistoric times. The early Forestry Commission was very much a State organisation with which most people had little interaction, and a fear of forest fires led to public access being discouraged, but the 1960s heralded the beginning of a great transition. Pioneering foresters and managers recognised that forests were more than a strategic timber asset; they were an environmental resource with huge potential for public recreation, as wealth and ease of access to the countryside grew. One of the leading proponents of change in the North York Moors was Gordon Simpson, a forester in the Moors from 1953 to 1993. He was amongst the first people to set public nature trails, and along with forest ranger Jack Eaton established a local history and natural history museum at Low Dalby – a forerunner to today's visitor centres.

A sometimes controversial aspect of the Forestry Commission's work was the visual impact on the landscape of the huge plantations – often of non-native conifers. Some efforts were made to mitigate this by planting hardwoods along roads, and where possible straight edges were avoided. The landscape architect Dame Sylvia Crowe was appointed to the Commission in 1963 as its first landscape consultant, and in the Moors she helped design the forest edge on Sneaton High Moor in order to soften its impact. Gradually, conservation and amenity gained increasing prominence in Forestry Commission policies.

For many people, the woods and forests of the North York Moors are now an important part of the landscape. They continue to provide a sustainable supply of timber and provide recreational opportunities for over a million visitors a year. The forest roads and streams support a huge network of grassland and riparian habitats and the wider forests are home to wide assemblage of life, from goldcrests to goshawks and butterwort to baneberry. Despite the regular threat of being wound up, the Forestry Commission has become very much an organisation of the people – and by 2017 tree cover in Britain stood at 13%, the highest for centuries.

Find out more

State-owned woods and forests in or near the North York Moors are:
Cropton Forest, Dalby Forest, Guisborough Forest, Langdale Forest, Newtondale Forest.

More information at https://www.forestry.gov.uk/visit

Langdale and Dalby Forests stretching into the distance, seen from above Silpho

Calcining kilns, Rosedale

1926

Rosedale and the Industrial Revolution

Linda Chambers

Tourists arriving in Rosedale 150 years ago (and there were a few even then) would have noticed a landscape of two parts. The upper dale-sides were covered in industrial installations resulting from the ironstone extraction and processing which was at its height, with a railway looping around Dale Head above the farms of the lower dale-sides, their fields full of crops and livestock. Miners poured in, along with railwaymen, attendant tradesmen and their families, from across the country and beyond. In the centre of this activity, Rosedale Abbey village was experiencing an almost total rebuild as significant amounts of money came into the dale for the first time in hundreds of years.

Mining was not exactly new to Rosedale. For many centuries, ironstone extraction had been carried out on a small scale. Excellent quality freestone provided blocks for buildings, and jet mining was a lucrative business during the mid 19th century, when pannier-loads taken by donkey to Whitby could make £500 for a few impecunious miners. But what was to happen in Rosedale through ironstone extraction was beyond the imagination of the community.

The first mines at Hollins (Low Works) at Rosedale West Side opened in 1856. The ironstone attracted big investors due to the unusually high content of iron, a geological anomaly which is still not fully understood.

George Leeman, the York MP and railway magnate, joined forces with Alexander Clunes Sherriff, a Worcestershire MP, and Isaac Hartas, owner of the local Wrelton ironworks. By 1861 the need for a railway to transport the most valuable ironstone in Britain had been addressed with the opening of an 11.5-mile stretch from Rosedale to the Ingleby Incline, running across a deserted moorland landscape. This line was built in 15 months through two winters by itinerant navvy gangs: the remains of their turf huts can still be seen near Dale Head. The connection onward to Battersby Junction and then to the processing plants of Cleveland and Durham allowed the partners to expand their business. By 1865 both sides of the dale were linked by rail, and mining activity increased due to world demand. Calcining (roasting) in three sets of gigantic kilns allowed transportation with less cost, and the high-grade ore made Rosedale an important player in the Industrial Revolution.

More than 100 miners' and railwaymen's cottages were built at the height of the boom. Miners' families formed their own communities on the exposed hillsides while farmers got on with their lives as they had for centuries. Many of the tenant farmers lived on the poverty line but in some cases, a son would find a job in the mines or on the railway, bringing in much-needed extra income.

Spare bed-space could accommodate a miner or maybe two or three, rotating (known as hot-bedding) as a shift ended.

This change and influx of people must have proved both daunting and exciting for the farming community. Benefits came from more schools and chapels, more money spent in the two inns, greater variety of goods in the shops, and a policeman. The behaviour of some of the incomers left something to be desired so police cells were opened for offenders to 'cool off' before appearing at the magistrates' court in Pickering.

Well over ten million tons of iron ore were transported across these bleak, often snow-covered, moors in the course of the railway's 68-year history. From periods of high production of armaments during the Franco-Prussian War in the 1870s, to the decline in the 1920s after the First World War, men and their families came and went in their hundreds. The population fluctuated accordingly.

It rose from 755 in 1851 to 3,024 in 1871, and reportedly a great deal more later in the 1870s. But in the 1920s the price of iron dropped, in 1929 the railway closed, and by 1931 the industrial age in Rosedale was past: the population was now just 498.

The visitors who walk, cycle and ride the railway line nowadays are often unaware of its industrial past and the importance of Rosedale's 'magnetic', high grade ironstone to the iron production of Cleveland and Durham. The remains of the giant kilns still dominate the landscape and the dale is still marked in many places by ruined dwellings and other structures to remind us of 70 years of mining. The remaining cottages make comfortable homes in a glorious setting. Farming continues, albeit on a reduced scale. Now the curlew and the ring ouzel fly unhindered through clear air, and rare plant species thrive in what was once a heavy industrial landscape.

Find out more

Former industrial sites can be visited on walks from Blakey Ridge and Rosedale Abbey.

Rosedale History Society, http://rosedale.ryedaleconnect.org.uk/history-contacts/

'Land of Iron' heritage project, http://www.northyorkmoors.org.uk/looking-after/landofiron

'Rosedale Mines and Railway' by R.H. Hayes and J.G. Rutter (1984)

1940

Crash of Heinkel Bomber near Whitby

Jane Ellis

At Sleights Road End, the junction of the A169 Whitby-Pickering road with the A171 Whitby-Guisborough road, is an unobtrusive stone pillar on which is an inscription by North Riding County Council (forerunner to North Yorkshire County Council). Few of the thousands of motorists who pass by it daily stop to read it. The pillar itself is interesting as it is a remnant of the bridge over the River Esk between Sleights and Briggswath which was washed away in a 1930 flood, while the plaque commemorates a significant event in our history, namely the first successful shooting down of an enemy aircraft on English soil in World War 2. It happened on 3rd February 1940, a bitterly cold and snowy day in the middle of a particularly severe winter.

At 9.03 am the operators at Danby Beacon Radar Station located three German medium bombers, Heinkel 111s, attacking an unarmed fishing trawler in the North Sea, killing the skipper and wounding two of the crew. Flight Lieutenant Peter Townsend was one of three fighter pilots from 43 Squadron scrambled from RAF Acklington, Northumberland, to intercept the Luftwaffe aircraft.

According to Townsend's 1969 book 'Duel of Eagles', since the cloudbase was low, he and his colleagues flew southwards over the sea at the maximum speed of their Hawker Hurricanes, almost skimming the waves so as to have the best chance of seeing the enemy before they themselves were spotted.

They engaged their adversaries at 9.40 am. Townsend was successful in disabling one of the Heinkels, no. 3232. Its German crew of four had been on duty since 2.00 am, helping to shovel snow and clear the airfield at Westerland, a remote airstrip on an island off the Danish coast, before they could take off for their five-hour flight across the North Sea. In the encounter observer Peter Leushake was killed outright, flight engineer Johann Meyer was fatally wounded, and Karl Missy, operating the top rear gun, was hit in his back and legs but continued to return fire - though his single machine gun was no match for the Hurricane's eight .303 Brownings.

Pilot Hermann Wilms, rather than ditch in the North Sea, decided to crash-land his damaged bomber close to Whitby, flying so low above the town that astonished local people could see him through the cockpit window. He guided his stricken plane down, narrowly avoiding hitting a barn but smashing though telegraph wires, and crash-landed near Bannial Flatt Farm. The plane's right wing hit a tree and it came to rest in a snow-covered field close to the farm buildings. The Whitby Gazette reported that the snow in the fields was stained red by a trail of the blood of the wounded airmen dripping from the aircraft as it approached the crash site. To this day there is a noticeable gap in the line of trees at this location.

Wilms managed to drag Leushake from the wreckage before realising that he was already dead, then with the help of Missy, who was himself badly injured, he managed to pull out Meyer, who had blood pouring from his wounds. Special Constable Arthur Barratt was one of the first on the scene and climbed into the cockpit to find Wilms burning confidential papers, though enough remained intact for the Intelligence Service to make good use of them. Whilst local people were attending to the injured, Wilms attempted to set fire to the plane, but with a combination of fire extinguishers and shovelfuls of snow the fire was put out. The two injured men, frozen with cold, were carried into the farmhouse, given tea and wrapped in blankets with hot water bottles. An ambulance took them to Whitby Cottage Hospital where Meyer was pronounced dead. Missy survived, though suffered a leg amputation as a result of his injuries. Townsend later visited him in Whitby Cottage Hospital; he shook Missy's hand and gave him a tin of 50 Players cigarettes and a bag of oranges.

The two airmen who did not survive were buried with full military honours at Catterick. The wreath on their coffins read simply 'From 43 Squadron, with Sympathy'.

Townsend, having risen to the rank of Group Captain, was to become well known in later years when he was romantically linked with Queen Elizabeth's sister Princess Margaret. Although the pair wished to marry, the 'Establishment' and the conventions of the time prevented them from doing so. In 1969 he was once again to meet Karl Missy, this time in the comradeship of ex-servicemen, and they remained friends until Missy's death in 1981.

Find out more

The memorial plaque is on a squat sandstone pillar near Bannial Flatt Farm,
on the south-east corner of the junction of the A171 and A169 just outside Whitby, grid ref. NZ 870 099.

More on the Heinkel Bomber: https://www.rafmuseum.org.uk/research/collections/heinkel-he111h-20/

More on the incident: http://www.yorkshire-aircraft.co.uk/aircraft/planes/40/1hfm.html

Bannial Flatt Farm cottages

N.R.C.C.
THE FIRST
ENEMY AIRCRAFT TO BE
SHOT DOWN IN ENGLAND
DURING THE
SECOND WORLD WAR
FELL 80 YDS. OPPOSITE
THIS TABLET
ON 3RD. FEBRUARY 1940

Plaque on the memorial pillar near the crash site

Model of the crash site, Whitby Museum

Frank Meadow Sutcliffe

Dock End, Whitby

Limpet gatherers

Girls shelling mussels

1941

Frank Meadow Sutcliffe, Photographer

Mike Shaw

Frank Meadow Sutcliffe was born in 1853 in Far Headingley, Leeds, the eldest son of Thomas Sutcliffe, a painter in watercolours whose qualities of good humour, excited curiosity and infectious enthusiasm Frank inherited. Thomas encouraged his children to make full use of their talents - they produced their own illustrated story books on his printing press - and when Frank showed an interest in photography Thomas cleared the hayloft to make a darkroom, and gave him a huge mahogany camera. Frank's formal schooling was almost non-existent but the stimulating company of his father, whom he sometimes accompanied on painting expeditions in the countryside, more than compensated. As a teenager he was already mastering the difficulties of the wet-plate process of photography and taking photographs of his friends and neighbours.

In 1870 the Sutcliffe family moved to Whitby, where they had spent their summer holidays since Frank's childhood. The following year Thomas died of heart failure at the age of 43, and Frank found himself head of the family. He was 18 and apart from an unhappy period as a clerk in Tetley's brewery in Leeds had no experience of making his way in the world. Before his death Thomas had suggested that Frank should take up photography as a profession, and with a sense of urgency he now began to transform an absorbing pastime into a means of earning a living and supporting his family.

To earn a living as a photographer in the 1870s almost invariably meant becoming a portrait photographer, including setting up a studio. A disastrous error of judgement led Sutcliffe to build a studio in Tunbridge Wells. This proved a total failure, and in 1876 he returned to Whitby with his wife and child. With the meagre resources he had left, he managed to rent part of a jet worker's shop; it was situated up a smelly back alley and became so hot on sunny days that sitters sometimes fainted. But it was a start, and Sutcliffe began to build up his business.

During the summer when Whitby came to life as an increasingly popular holiday resort, Sutcliffe worked in his studio from early morning to late at night, and when he went home continued mounting his photographs into the early hours of the morning. But he soon realized that he could not earn enough in the six weeks or so of the season to carry him through the rest of the year. The spur of financial necessity added to the pleasure of photographing his surroundings, and it was through reasoned desperation - as a friend later described it - that Sutcliffe began to take the photographs which were to make him famous.

His photographs are striking visual images which reveal his great skill as a photographer, but more important is the sense of deep affection which reaches the spectator. Sutcliffe delighted in a love for Whitby, for Eskdale and it inhabitants, and for the surrounding moorland. His photographs reveal a close personal attachment to his subject matter. He cared about the people he was photographing - the fishwives, the fishermen, and the farmworkers.

In 1894 he moved into a new studio in Skinner Street, Whitby. His business as a portrait photographer was flourishing, and he found less time to spend on the photographs which by then had brought him worldwide fame. His work had been exhibited in London in the first one-man show to be held by the Camera Club in 1888; the Prince of Wales had purchased a print of 'Water Rats' - probably Sutcliffe's most famous photograph - and between 1881 and 1905 his photographs were honoured with 62 gold, silver and bronze medals at exhibitions in Britain and all over the world, from Chicago to Tokyo and from Paris to Calcutta. He now put a good deal of energy into magazine and newspaper articles on photography. His excitement with the images produced by photography continued unabated, and around the turn of the century he became absorbed in working with the new range of hand-held Kodak cameras.

When in 1922 he sold his photographic business, his retirement lasted just one week, after which he became curator of Whitby Museum. In 1935, six years before he died in 1941, he was made an Honorary Fellow of the Royal Photographic Society. It was a belated recognition of the talents of a photographer who set high standards for himself and, by treating photography as a creative medium in its own right and not as a debased form of painting, produced some of the most striking yet unpretentious photographs of the 19th century.

Find out more

Sutcliffe Gallery, 1 Flowergate, Whitby YO21 3BA,
http://www.sutcliffe-gallery.co.uk/ (with online gallery of Sutcliffe's photographs)

http://www.amateurphotographer.co.uk/iconic-images/frank-meadow-sutcliffe-1853-1941-iconic-photographer-15171

1947

The 'Long Winter' in the North Yorkshire Moors

Carol Wilson

The winter of 1947 is remembered by all who lived through it. There had been some cold weather in December 1946 but this was followed by an unseasonably mild spell. However, on 22nd January 1947 the temperature dropped like a stone and the snow began in earnest. The severe weather was not to relent until mid-March: from 22nd January until 17th March snow fell every day somewhere in the UK.

Although more snow was to fall during the winter of 1962/3, that of 1947 was much colder, with biting winds that drove the snow. Drifts were up to seven metres deep while, at its lowest, the temperature plummeted to -21°C. There was also very little sunshine, with only 17 hours recorded at Kew for the whole of February.

The drifting snow caused widespread transport difficulties: 300 main roads became unusable, and over 100,000 British and Polish troops, as well as German prisoners-of-war, were drafted in to clear the railways as the disruption was seriously affecting the coal supply to power stations. Domestic power supplies were reduced to just 19 hours per day, which continued until the end of April. Industrial supplies were cut completely and all external lighting was switched off. Radio broadcasting was limited and the television network suspended altogether.

Flocks of sheep and herds of cattle were buried and froze or starved to death. Nationwide, as much as a quarter of the sheep stock was lost and it took farmers six years to recover from such losses. Vegetables became frozen in the ground. War-time rationing was still in place and the winter of '47 saw even more stringent measures to eke out what was available.

In the immediate aftermath of World War 2, the UK was facing serious economic difficulties. Britain was still involved in post-war defences which drew considerably on the public purse. Added to this, the Labour government had undertaken a substantial programme of nationalisation, including the railways and the coal-mining industry. Alongside all of this was the need to fund the new National Health Service. There was not a lot to spare. It was this that caused as much hardship as - if not more than – the snowfall.

On the North Yorkshire Moors, where the wind-chill factor can be appreciable in any winter, the snow filled the roads and lanes, cut off villages and caused considerable difficulties for farmers. The loss of the electricity supply, however, made no difference as electricity supplied by the National Grid only reached here in the early 1950s. Instead, folks simply soldiered on with their paraffin lamps and stoves and used peat on the fire or in the range for cooking. For those who needed coal for heating, when paraffin refills were necessary or when food supplies began to run low, people acted as in Westerdale, where they hitched their horses to sledges and, with the lanes completely blocked by snow, made their way across the fields to Castleton. Coal could be obtained from the railway station here and limited food supplies were available in local shops - but it took a long time to get there and even longer to get home.

Newspaper headlines declared that Scotland was 'cut off' and England was 'cut in half'. Villages became completely separated from one another and remote farms entirely isolated. Moorland folks, however, were used to hard work and harsh conditions and simply ploughed on with life. Largely self-sufficient – and certainly self-supporting – the communities of the North Yorkshire Moors managed as best they could. Used to helping each other, everyone rallied together and made the best of it.

Sadly, even once the snow abated, worse was to follow as the mid-March thaw and additional rain caused widespread flooding. It was the wettest March for 300 years. Thirty-one counties across Britain and at least 100,000 properties were affected by the floodwater, and the army was once again called in to help. Red Cross parcels arrived from Australia and Canada and it took ten days for the flood waters to subside.

Throughout the difficulties, over two million people were forced to claim unemployment benefit because of the loss of income, but there was little unrest and no major public disorder. However, following such hardship some decided to strike out for a different life and many people emigrated, in particular to Canada and Australia. In Britain, taxes were raised once more - but the country then basked in a glorious warm summer.

(With grateful thanks to the late Emma Beeforth of Westerdale for help in preparing this article.)

Find out more

Blog posting from the Met Office -
https://blog.metoffice.gov.uk/2017/01/26/winter-1947-brought-a-freeze-to-post-war-britain/

Rare film footage of a family's experience of the 1947 winter in West Yorkshire -
https://www.youtube.com/watch?v=rUcAfHD1BAY

Sid Beeforth and god-daughter Audrey Shaw, Westerdale

The Stothard brothers, Mount Pleasant Farm, near Lockton

Ivy Holme, Westerdale

Feeding sheep at Mount Pleasant Farm, near Lockton

Rosedale after the blizzards

Ralph Cross

Boundary marker stone above Rosedale

National Park boundary sign

Red Grouse, synonymous with the Moors

1952

Establishment of the North York Moors National Park

Janet Cochrane

In medieval times, forests and mountains were seen as dangerous places where wild beasts and demons roamed. Myths and legends reflecting such fears persisted for centuries and appear in place names across the North Yorkshire such as Boggle Hole, near Whitby – believed to be where the 'little people', or 'boggles', lived - and 'Fairy Cross Plain' in Fryupdale, the home of elves and fairies.

By the end of the 18th century such beliefs were disappearing as understanding of the natural world evolved. In the industrialised world people were increasingly concentrated in cities, with a corresponding shift in perceptions of nature: the Romantic movement expressed these by harking back to a supposedly more natural and purer past. The philosopher Jean-Jacques Rousseau eulogised the mountains and peasantry of his native Switzerland, and in Britain the poems of William Wordsworth created awareness of the beauty of wild places. Even though the Romantics' ideal of a medievalist, rural idyll had never actually existed, they influenced the European educated classes to consider nature in a more positive light.

In America, the Romantics seized on the wilderness as a source of inspiration. Artists were employed by the railway companies to paint dramatic landscapes which would encourage urban-dwellers to visit these areas (travelling, of course, by train). In North Yorkshire, a similar link between the railways, tourism and the scenery was created by the Staithes Group of artists, who worked in villages along the northeast coast in the late 19th century. They sold their works to the emerging middle-classes of the industrial cities, stimulating them to visit and appreciate unpolluted and more natural scenery.

All this helped to win popular support for landscape preservation, bolstered by the Victorians' enthusiastic study of natural history. There were increasing demands for greater countryside access after World War 1 and finally, after the Second World War, 10 national parks in England and Wales were designated under the 1949 National Parks and Access to the Countryside Act. The North York Moors was one of these, formally coming into existence in 1952.

The National Park measures 143,603 hectares and consists of a high sandstone plateau pierced by valleys of woods, rivers and farmland. From the Vale of Pickering in the south the land gradually rises to the high moorland and then falls away to the north in an escarpment, facing Teesside. The principal valleys such as Bilsdale, Farndale and Rosedale run north-south and drain southwards into the River Rye and thence to the Derwent, the Ouse and

99

the Humber, while to the north the valleys drain into the Esk which flows to the sea at Whitby, and in the north-west the Leven drains into the Tees.

The North York Moors boasts the greatest expanse of heather moorland in England and Wales, but 5,000 years ago both uplands and valleys were wooded. These were gradually cleared as human settlements spread uphill during the Bronze Age (from about 2100 BC). Increasing agricultural activity during the Romano-British period saw further clearances, and again during the Anglo-Scandinavian era – but woodland expanded again when William the Conqueror's 'harrying of the north' in AD 1069-70 caused famine and depopulation.

Monasteries such as Rievaulx, Byland and Mount Grace revived the agricultural economy in the 12th-14th centuries and led to the management of moorland for sheep to supply the lucrative trade in wool. Nowadays moorland management balances sheep production with habitats for grouse and other wildlife, contributing both to conservation and to the local economy. A wide range of tourism, agricultural, craft and other enterprises reliant on the resources of the Moors helps to support a population of around 24,000 people living in over 90 villages and scores of isolated farms, creating the thriving social-ecological system we see today.

For visitors, the Park offers a huge choice of activities, from adventure and activity sports such as mountain-biking, rock-climbing and horse-riding to visits to ruined castles and abbeys, ancient inns and rides on the heritage North York Moors railway, or opportunities just to sit and admire the coastal views at Runswick Bay or Robin Hood's Bay, or inland at Sutton Bank or Hole of Horcum. There are any number of short walks, and longer walks which are completely or partly within the national park include the Coast-to-Coast Walk (192 miles), Cleveland Way (109 miles), Lyke Wake Walk (40 miles) and the Esk Valley Walk (37 miles) – not forgetting the 'special interest' Inn Way, covering 89 miles and 31 pubs!

Find out more

North York Moors National Park Authority,
http://www.northyorkmoors.org.uk/

North Yorkshire Moors Association,
http://www.north-yorkshire-moors.org.uk/

1953

Coronation Celebrations in the Esk Valley

Tamsyn Naylor

On June 2nd 1953, the Coronation of Elizabeth II took place at Westminster Abbey in London. She was 25 years of age, the 39th monarch to be crowned at the Abbey, the setting for British coronations since 1066. Millions were able to watch the proceedings televised for the first time and it was a day of rejoicing and celebration across the United Kingdom, the Commonwealth, and the wider world. Local celebrations, presentations, parades and street parties took place across North Yorkshire including the Esk Valley and Danby area, where the History Tree had been standing for some 150 years, through the reigns of eight previous monarchs.

One thing the History Tree would have liked about the Coronation was the relentless rain that came with it! June 2nd was a notable day, not least due to the 50-mile an hour storm that rained on any planned local parades. But the spirit that made Britain great was not to be defeated and the celebrations went ahead.

A reporter from the Whitby Gazette toured the villages of the Esk Valley, stopping first at Egton. Coronation decorations festooning the village received a severe buffeting from the storm, but freshly out of a church service and now tucked up warm in the schoolroom, partaking of a sumptuous tea, were the parishioners. The assembled children, adorned for the fancy dress parade, were taken by coach to the Tenants' Room at Egton Bridge, along with the decorated tractors, perambulators and bicycles, where judging took place. All 144 village children were presented with souvenir mugs and the day was rounded off with a memorable social evening.

The next stop was Glaisdale, where the village 'made merry' in the Robinson Institute. The sports activities were postponed and the children in fancy dress were in danger of being disappointed, until the decision to bring them and the decorated wheeled vehicles inside for judging was taken. The abandonment of the parade from the railway station was accepted philosophically, leaving outside the Master of the Glaisdale Hunt, Jim Winspear, who had arrived on his steed Robin Adair dressed as Winston Churchill, complete with authentic cigar – a brown paper-wrapped skittle. The ample tea was served with Coronation cake, and 156 souvenir mugs were presented to the children. The beacon lighting above Hall Farm was put off till Friday.

At Lealholm the fancy dress competition took place in the schoolroom, with 80 entrants, followed by ice-cream. Tea for children and parents was supplied in the Shepherd's Hall, with the elderly residents served in

Nelson Hall, followed by a social evening. Three days later the Maypole dancing took place, when the sun blazed down from a cloudless sky and a colourful procession made its way from the village green and over the bridge. The sports were topped with a tug-of-war between married and single women, and after a great struggle the singletons were declared victors. A huge bonfire was lit, followed by a fireworks display.

Other villages celebrated in similar style. At Danby, prizes of Coronation souvenirs were given not only for costumes and decorated cycles but also for decorated houses and shop window displays. The day ended with a huge bonfire on the Howe and the discharge of colourful Coronation fireworks. Souvenirs in Fryup included walking sticks, tobacco, and tins of tea or chocolates with the Queen's portrait. Grosmont Coronation Committee hired a car to take the infirm folk to the Methodist schoolroom, where a television had been installed to show the ceremony, and the village jazz band played dressed in Union colours and kilts to cheer the drabness of the evening.

Goathland delayed the planting of a commemorative tree and installation of a bench but adorned tea-tables in the parish hall with flowers, flags and balloons. The children were presented with mugs, ice cream and an orange, while the local Country Dance Club gave a dancing display. Villagers at Littlebeck laid out a huge Union Flag measuring 30 feet by 16 feet from crepe paper once the weather improved, and gardens looked splendid planted in Union colours.

One beneficial spin-off from the Coronation was the distribution of thousands of acorns from mature oaks in Windsor Great Park to places around Britain and to Commonwealth countries, to be planted in parks, school grounds, cemeteries, private estates and gardens. These trees became known as Royal Oaks or Coronation Oaks and have added to the robustness and beauty of our countryside.

In 2017 the Queen became the first British monarch to celebrate a Sapphire Jubilee, commemorating 65 years on the throne. A remarkable achievement!

Find out more

A personal account of the Coronation -
http://www.historic-uk.com/HistoryUK/HistoryofBritain/The-Coronation-1953/

Fascinating glimpses of 1950s life and times in Coronation celebrations in a Yorkshire village -
https://www.youtube.com/watch?v=0HHzw5cJcPM

Coverage of the Coronation: https://www.youtube.com/watch?v=k3HrsLeZJ_E

A celebratory Coronation mug

Mount Everest ('Chomolungma' is its Tibetan name)

Adele Pennington at the Everest Summit

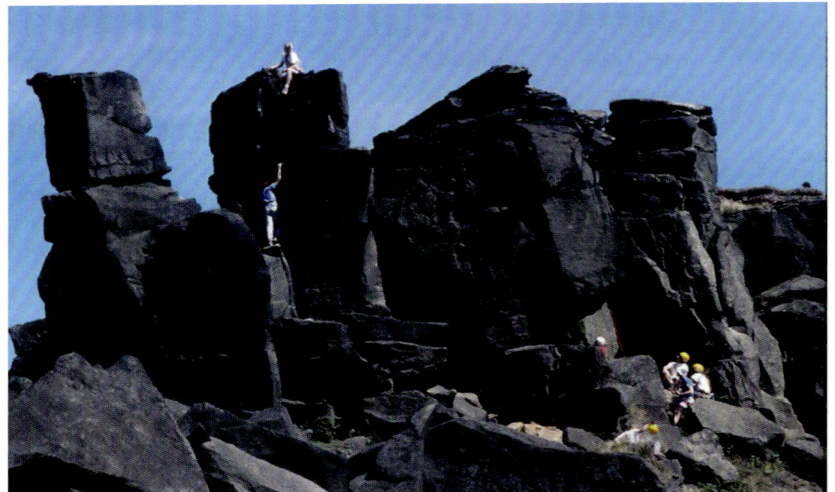

Climbers on the Wainstones, Hasty Bank

The Summit of Mount Everest is Reached

Adele Pennington

Over 4500 miles from Mount Everest lies the highest point of the North York Moors National Park, Round Hill on Urra Moor, at just 454m above sea-level. Although it is marked by a trig point and stands close to the Cleveland Way long-distance walk, many would undoubtedly reach this point not realising they had reached the summit of this mass expanse of beautiful moorland.

On 29th May 1953, by contrast, there was no doubt that Tenzing Norgay Sherpa and Edmund Hilary were on the top of Mount Everest (8848m) - the first two people to reach it – and the news travelled to England just in time for Queen Elizabeth II's Coronation. Hilary and Tenzing were part of a British expedition led by John Hunt, a well-known British mountaineer and a colonel in the British Army.

An iconic landmark and more mountain-like summit in the North York Moors than Round Hill is Roseberry Topping, at 320m – Everest is almost 28 times this height! This local 'mountain' lies close to the village of Great Ayton, on the northern flanks of the national park, and thousands trek each year to its peak to take in the magnificent views. It is often referred to as the 'Matterhorn of the Moors' as it resembles the shape of the famous mountain in Switzerland. There is no definite date when the first person summited the distinctive Roseberry, but it is well known that James Cook as a young boy regularly took himself up the hill and it was perhaps this that gave him his taste for adventure and exploration.

Despite the lack of any major cliffs or mountains on the Moors, local climbers started to explore the crags that bound the upland plateau in the 1920s. Many famous climbers honed their skills on these crags before making their historic names in the timeline of British, Alpine and Himalayan climbing. Throughout the last century, the modest crags of the North Yorkshire Moors, with names such as Scugdale, the Wainstones, Ravenscar, Highcliffe Nab, Park Nab and many more have provided the rock-climber with numerous challenging problems.

In the same year that Everest was conquered the Cleveland Mountaineering Club was founded. Now with over 150 people on its membership list, club members enjoy summer days and evenings rock-climbing and exploring the wild expanse of the North Yorkshire Moors. The climbing guidebook to the area was written by the club's members and has lured many climbers to the crags. As the climbing grades have risen over the years, talented climbers are still pursuing new routes to tantalize others and entice them to take up the challenge of stretching their abilities..

Alan Hinkes OBE, the first Briton to climb all of the world's 8000m peaks (there are 14 of them) is a fellow member of this club, and is still to be seen climbing and walking in the area. Inspired by his conquests and a dream to climb Everest, I pursued a training routine of running up Roseberry Topping and soloing many easy routes at Scugdale after teaching Chemistry all day at a local secondary school. Older members of the club passed on their experience and wisdom and smiled at my antics, practising crevasse rescue off the edge of Captain Cook's crags and fixing my own line to ascend steeper routes. For me this was a way to gain confidence and skills and make small steps towards an as-yet unfulfilled dream.

In 1998, I was lucky enough to be invited to join a Himalayan expedition to make a winter ascent of Ama Dablam (6836m) - the 'Matterhorn of the Himalayas'. I was the first British woman to do so, and the view of Mount Everest was overwhelming. The sharp contrast of the Himalayan mountains with snowcapped peaks, as opposed to the blanket of heather that was familiar from the North York Moors, led me to take evening jaunts to the higher hills and crags of the Lake District and to spend weekends in the Scottish Highlands at every opportunity. As with the early pioneers of Everest, the self-indulgence of climbing and training took over and in 2008, 55 years after the first ascent, a dream came true: I reached the summit of Mount Everest.

Now, a further ten years on and with two summits of Everest and a further five 8000m peaks, the opportunity to walk across the moors, climb at Scugdale and summit Roseberry Topping is nothing but pure pleasure.

Find out more

Cleveland Mountaineering Club,
http://www.clevelandmc.com/

Rock-climbing in the North York Moors,
http://www.climbonline.co.uk/north_york_moors1.htm

BBC News report of the first ascent of Everest,
http://news.bbc.co.uk/onthisday/hi/dates/stories/may/29/newsid_2492000/2492683.stm

1955

Camphill Village Trust - Botton Village

Rita Leaman

Nestling in the North York Moors is the village of Botton. This is no ordinary village. It is a special village with a special community: part of the Camphill Village Trust.

The dream of such a community began with a group of likeminded, mostly Jewish, Austrian refugees, who fled their homeland in 1939 and arrived in Aberdeen. Their aim was to set up a therapeutic community for children with learning disabilities and they found a sympathetic landowner willing to rent them a suitable property. Unfortunately, at the outbreak of war, the men were interned as enemy aliens for six months, but the women continued with the work. The men returned and the Camphill Movement started in earnest. On June 1st 1940 the 'Camphill Community for Children in Need of Special Care', was opened on the Camphill Estate in Scotland, given to the Camphill Movement by the Macmillan family. A loan of £1000 was also given from the Scottish Council for Refugees.

The natural leader of the Camphill Movement was Dr Karl König, who took his inspiration from the Austrian philosopher, Rudolph Steiner, founder of the anthroposophy movement: this posits that the spiritual world can be accessed by direct experience through inner development.

Dr König studied Medicine at the University of Vienna and graduated in 1927 with a special interest in Embryology. After graduating, he met Ita Wegman, an anthroposophical physician who invited him to work in her institute for people with special needs in Arlesheim, Switzerland. Following his work in Arlesheim, König was appointed paediatrician at the Rudolf Steiner inspired Pilgrimshain Institute in Strzegom, where he worked until 1936, when he returned to Vienna and set up a successful medical practice. In Aberdeen, Dr König was further inspired to help children with learning disabilities, because he felt they were treated as outcasts in their communities, as he and his colleagues had been in Austria.

In the first year, nineteen pupils from the ages of two to nineteen were educated in therapeutic groups. Dr König called the work 'Curative Education'. This was based on the anthroposophical approach, that 'a perfectly formed spirit and destiny belong to each human being'. The success of the venture led to expansion of the community through the 1940s and by 1950, there were 222 pupils at Camphill schools, with another 156 on the waiting list. Official Ministry reports showed that these schools were then the only form of educational provision in the UK available for children with severe learning disabilities. The work expanded to assist children with a variety of special needs.

In October 1954, the Camphill Village Trust Ltd was set up and in 1955, with an increasing number of requests for communities for adults, the first Camphill Village Trust Community was founded in Botton village, in Danby Dale, North Yorkshire. By 2017 there were nine Camphill Village Trust Communities in the UK, providing homes for around 400 adults with learning disabilities, mental health problems and special needs. Worldwide, the Camphill Movement has established over a hundred therapeutic communities in more than twenty countries.

Botton Village is home to 150 people, 90 of whom have a learning disability and other special needs. The village is spread over 650 acres of land and supports five biodynamic farms and a number of workshops and gardens, providing a variety of working experiences. The accommodation, which is shared with co-workers, comprises thirty houses ranging from farmhouses and barn conversions to purpose-built accommodation and houses for the elderly.

With the emphasis on community, the village also has a village hall for shows and gatherings, a church, village shop, organic bakery, health centre, post office, and coffee bar where visitors are welcome to join the residents. A gift shop sells items made by the residents in the workshops and from other Camphill communities. There is also a Waldorf Steiner school, educating children from Botton and surrounding areas.

The ethos of the community is for each person to be treated as an individual with their own individual needs, being valued and appreciated. These needs are met through training and work, learning new skills, enjoying social interaction and friendship, together with cultural and spiritual guidance.

Botton village received the Deputy Prime Minister's Award for Sustainable Communities in 2005; the award cited the community's dedication to sustainability and mutual respect and their achievements in these areas. The judges said, "Botton offers both lessons and inspirations to the endeavour to create sustainable communities. The community's culture focuses on tolerance, respect and developing individual strengths."

Find out more

The Camphill Village Trust and Botton Village,
https://www.cvt.org.uk/communities/botton-village

History of the Camphill Movement,
http://www.camphill.org.uk/about/camphill-history

Botton Hall, Camphill Village Trust, in the head of Danby Dale

Lilla Cross, Fylingdales Moor

1955

Lilla Cross on Fylingdales Moor

Albert Elliot

The moorland areas of northeast Yorkshire are studded with the remains of three dozen or so named crosses or waymarker stones, said to be Britain's largest concentration of old stone crosses. Due to the crosses' great antiquity, it is impossible to know with any certainty by whom or for what purpose they were placed there, but most were probably erected in the Middle Ages for use as boundary markers or as waymarkers to direct travellers safely along their route. The crosses seem to sit comfortably in their surroundings and some in particular have an immutable air of mystery attached to them: Lilla Cross is one of them.

These stone relics of varying ages exist in differing conditions. Some have only their socket stones surviving. Many are now just stumps with shafts either totally or partially missing, some have undergone repairs after suffering natural wear and tear or have been damaged by wanton vandalism. Others are, miraculously, still standing intact and in remarkably good condition considering their great age and the constant battering and attrition they have endured from exposure to the harsh elements. Perhaps the best known of the moorland crosses is Young Ralph Cross on Blakey Rigg, adopted in 1974 as the main emblem on the official logo of the North York Moors National Park, following a public competition.

Lilla Cross on Fylingdales Moor has survived down the centuries in remarkable condition. Thought to date from the 10th century, Lilla Cross is the oldest on the North Yorkshire Moors and regarded as one of the earliest Christian monuments in northern England. It is a protected historic monument.

Many people ask about the origin of the name. Legend has it that Lilla was the faithful minister or man-servant of Edwin, King of Deira and Bernica (later known as Northumbria). A rival king of the West Saxons sent an assassin to kill Edwin. On Easter eve in the year 626 the assassin made an attempt on the King's life by lunging at him with his double-edged poisoned dagger. Lilla, who was nearby, anticipated what was happening and swiftly flung himself between the King and the assassin's blade. In doing so Lilla was fatally wounded and died on the spot. The thwarted assassin was slaughtered by the King's men. The King's life was saved and he was so moved by the bravery of Lilla, a Christian, that in recognition of his loyal servant's act he ordered a stone cross be erected at the very place where the murderous attempt took place. The king also commanded that the dead hero be buried beneath it. Although the current monument dates from many years after the event it commemorates, it is known as Lilla Cross.

In 1952, Fylingdales Moor became a military training area, and Lilla Cross was moved from its ancient site and repositioned at Simon Howe, well away from harm. When some years later military activity ceased in the area, it was decided to return the monument to Lilla Howe, its original location. This was a delicate job as the cross had unfortunately been set in concrete when moved to its temporary location. Fortuitously, the work was to be carried out by a group of Sappers (as the Royal Engineers are popularly known) of 508 Field Squadron from Horden, a colliery village on the coast about five miles north of Hartlepool, County Durham.

The first task facing the soldiers was to construct a special supporting cradle in which the Cross could be safely transported to the Squadron Drill Hall at Horden, where the difficult task of removing the one and half tons or so of unwanted concrete could be undertaken. It was of great advantage to the project that the Sappers involved were experienced miners in daily life and expert with hammer, pick and chisel; they had just the skills required for such a delicate task. They went on to complete the work and re-erect the Cross without harming the 1300-year-old relic: quite an achievement. As a testament to the Sappers' work, a plaque is sited nearby the ancient monument.

Today the ancient cross still stands erect on Fylingdales Moor in tribute to Lilla, King Edwin's faithful servant who saved his master from certain death by the assassin's knife. Lilla Howe is a very atmospheric place. Go alone on a quiet evening and stand and listen by the lonely Lilla Cross on the moor and you can almost hear the cries echo down the years from that fateful day over thirteen centuries ago: history comes alive.

Find out more

Historic England on Lilla Cross and others:
https://historicengland.org.uk/listing/the-list/list-entry/1010076

Steve Estill's 2017 book, 'Stones and Crosses of the North York Moors', published by Fonthill Media, is an excellent guide to Lilla Cross and other stones.

1964

North Skelton, the Last Ironstone Mine in East Cleveland

Beth Andrews

At noon on 17th January 1964, miners finished their last shift underground at the North Skelton ironstone mine. They followed the ore they had blasted and collected to the surface for the last time, as the mine fell silent. As the final load of ore from the Cleveland iron ore field ran along the railways to the blast furnaces of Teesside, it was the end of the Ironstone Era of East Cleveland.

Legend has it that on 8th June 1850, John Vaughan and Henry Bolckow were looking for a local source of ironstone to send to their blast furnace in the new town of Middlesbrough. Encouraged by the opening of the drift mine in Skinningrove in 1848, they decided that Cleveland ironstone was the obvious choice. Walking along the Eston Hills with mining engineer John Marley, Vaughan stumbled in a rabbit hole, and on examining a stone nearby he realised that this was the ore he was looking for. Further research showed that this was the main seam of the Cleveland Ironstone Formation, which at Eston is 4.8 metres thick. Eight weeks later, ironstone was being quarried and sent to furnaces across the area.

The massive and growing demand for iron to feed the Industrial Revolution drove ironmasters and landowners to open mines wherever ironstone was found. The iron-rush of East Cleveland had started, and with it the growth of Middlesbrough, which became known as 'Ironopolis', one of the most important iron-producing districts in the world.

Early mines were set up where ironstone was found at the surface and then the seam of ore was followed underground via drift mines. Soon, shaft mines joined the expanding number of drift mines as the railway made its way across East Cleveland.

The ironstone seams in the Cleveland area are not flat, having been folded and faulted over time, so finding ironstone was not always easy. The ore was often hidden deep below the surface and in some places had been displaced by faults. The ironstone seams in the Skelton area form a saucer shape and the ore lies at over 220m deep. Extracting stone from these deep mines is difficult; not only was there a need to get people and machinery to the bottom of the mine and then lift the stone to the surface, it was also essential to deal with the ingress of water and prevent flooding. North Skelton mine was 120m below sea-level, and extracting the iron ore required many drains and pumps to keep the working areas free of water.

Development of the mine began in 1865 with the creation of a water-level near Marske Mill that was to help drain the mine in the future. This part of the workings became known as North Skelton. As understanding and knowledge of ironstone in the area developed, the mine shaft was sunk at a point further south. However, the original name stuck and that area too became known as North Skelton - despite being located south of the village of Skelton.

North Skelton Mine, commissioned in 1875, was the deepest mine in Cleveland, with a 230m shaft. Despite the long period of time and the investment needed to get started, the mine was one of the most successful in East Cleveland. At its peak - between 1875 and 1920 - nearly 6 million tonnes of ore were taken out of the ground annually. Overall, 360 million tonnes of ore were extracted during the 113 years of industrial operation.

As with most of the ironstone mines, North Skelton did not have a local population large enough to provide workers, and a new village was built to house the miners and their families who came from all corners of Britain and beyond. Often built by the owners of the mines, these villages had rows of terraced houses with allotments, shops, schools and churches. Whole families and communities worked in or depended on the mines for their livelihoods. However, mining was a difficult and dangerous occupation and injuries and fatalities were common. Falling stones, blasts and injuries from hauling wagons could easily prove fatal.

With the import of cheaper ore and discovery of higher quality ore around the world, one by one the mines began to close. With the closure of North Skelton mine and removal of most of the buildings, headgear, spoil-tips and railways, the villages and their communities are all that remain as reminders of this once great industry which changed the landscape of East Cleveland and created modern Teesside.

Find out more

Interactive website on Ironstone Mining in Cleveland,
https://www.redcar-cleveland.gov.uk/kirkleatham/collection/cleveland-ironstone.pdf

Cleveland Ironstone Mining Museum, Deepdale, Mill Lane, Skinningrove TS13 4AP
http://ironstonemuseum.co.uk/

Derelict mine buildings in front of a spoil heap, North Skelton

Old shops, with Tudor Manor House from Harome in the background, Ryedale Folk Museum, Hutton-le-Hole

1964

Ryedale Folk Museum: In the Beginning

Elaine Wisdom

Ryedale Folk Museum officially opened in 1964, but its antecedents lie over thirty years earlier. The founding figures are credited as Bertram 'Bert' Frank, Wilfred Crosland and Raymond Hayes, but others played their part too, notably Wilfred's sisters, Helen and Hannah (known as Minnie), and Bert's wife Evelyn.

In 1935 Wilfred Crosland and his sisters were living at Elphfield in Hutton-le-Hole. To display his collection of artefacts found on the Moors and raise funds for the Village Hall, they opened one room of their home for two weeks. Wilfred had collected for much of his life and dreamed of his collection forming a local museum. The outbreak of World War 2 put paid to this, but during the late 1940's Crosland and Bert Frank, then living at Lastingham, met to discuss the possibility of a permanent collection. No progress was made, however, and in 1961 Wilfred Crosland died.

Bert also spent much of his life collecting on the Moors, and from 1960 housed his collection first in a garden shed and then in four small buildings, with the support of enthusiastic volunteers including Raymond Hayes, a local archaeologist.

Efforts to find space for the fledgling museum to expand were fruitless, and Bert felt the venture was doomed to fail: "…only a miracle could save it….". But that miracle occurred when in July 1962 Bert was invited to meet Wilfred's sisters, Helen and Minnie. His diary records: "I felt that I was going to have an extraordinary experience … They had heard that the infant museum which I had started at Lastingham … was in danger of collapsing due to lack of space. They went on to say that … if I was willing to leave Lastingham and come to live at Hutton-le-Hole, I could take over the remnants of their brother's collection". The Crosland sisters said they would leave the house and its grounds to Bert in their will to turn into a museum.

Fortunately, Evelyn Frank agreed to the move, and in November 1962 Bert and Evelyn cleared and whitewashed the first room of the new museum, where Wilfred had kept his own collection. The one room soon expanded to three, by August 1963 a board bearing the legend 'Ryedale Folk Museum' was attached to the outside wall, and the doors opened for the first time to the public after a small ceremony.

Bert's diary for 5th August 1963 reads: "Museum … was visited by a number of people who paid one shilling to see the exhibits … We were surprised to find how many local people were interested enough to come and see the collection for themselves; some of them bringing household objects, craftsman's tools, and other implements". 1964 saw Bert clearing out the barn for additional space; this area is now the entrance to the present museum, the striking arch being added in 1994.

Minnie and Helen Crosland did not long survive the museum's opening, dying in 1963 and 1964. Bert and Evelyn continued to live at Elphfield and the museum officially opened on 28th March 1964, with the Look North TV crew visiting in April. The work of collecting continued, and in 1965 the first vernacular building was offered to the museum. 'Stang End' was "an old thatched cottage of cruck construction…". It was dismantled and reconstructed in the grounds of the museum. A unique acquisition was the glass furnace from Rosedale, discovered by Raymond Hayes and brought to the museum after much hard labour and ingenious ideas for its removal from boggy ground. Constructed in 1570, it had remained in use until about 1600.

In 1970 Harome Manor House, built around 1570 and home to the Lords of the Manor and the Manor Courts, was offered to the museum. During its dismantling, a silver spoon of 1510 was found in the thatch, then considered the oldest spoon of its type in the country. The British Museum now has it in its collection, but it made a welcome return to the museum in 2017 as part of an exhibition. In 2012 the museum was further augmented by the extraordinary Harrison Collection of antiques and curiosities from 500 years of British history, collected by brothers Edward and Richard Harrison over five decades.

Bert Frank died in 1996, having seen the dreams of Wilfred Crosland and himself become a reality. The museum is now a major tourist attraction in the North York Moors, recording and commemorating a way of life now gone or fast disappearing.

Raymond Hayes and Bert Frank were awarded MBEs in 1986 in recognition of their services to the archaeology of the North York Moors and to the foundation of Ryedale Folk Museum respectively.

Find out more

Ryedale Folk Museum, Hutton-le-Hole, Kirkbymoorside, YO62 6UA,
https://www.ryedalefolkmuseum.co.uk/

Yorkshire Vernacular Buildings Study Group,
http://www.yvbsg.org.uk/

1968

Sir Herbert Read

Colin Speakman

Herbert Read was one of the most influential British art historians, critics, essayists, educationalists, philosophers and poets of the 20th century. Born in 1893 at Muscoates Grange Farm, four miles south of Kirkbymoorside, Herbert enjoyed an idyllic childhood, beautifully and vividly described in his autobiographical 'Annals of Innocence and Experience'.

But things changed dramatically when in 1902 his father died and the farm had to be sold. Herbert was sent away to a boarding school near Halifax, where his only solace was the joy of reading. He left school at 15 for a job in a Leeds Bank and, largely through night school and the public library, passed matriculation exams. Thanks to a small legacy, he went to Leeds University to study law and economics, despite having a greater interest in arts, politics and poetry. Immediately after he finished university in 1914, war broke out.

Herbert joined the Green Howards and served as an officer, enjoying a career in the infantry which earned him both a Distinguished Service Medal and Military Cross. The horror of war – the 'Annals' contains a moving description of his encounter with a German officer who was also a learned academic – transformed him as a writer, and he began to produce a series of powerful poems that established him as one of the leading War Poets.

Civilian life drew him to London and a Civil Service job, but his essays and articles in a variety of literary and arts magazines soon established his reputation as an expert on the visual arts. Part of a highly influential intellectual circle, he became among the first to recognise such major figures as T.S. Eliot, Henry Moore, Ben Nicholson and Barbara Hepworth. He was for a long period a curator of the Victoria and Albert Museum, then Professor of Fine Art at Edinburgh, editor of the Burlington Magazine, and co-founder of the Institute of Contemporary Arts.

Among his many books and publications, 'The Meaning of Art' and 'Education through Art' went through several editions and had a huge influence on the teaching of arts and aesthetics by a whole generation of teachers and writers. Deeply influenced by the great Romantics, notably Wordsworth, Blake and Coleridge, Herbert Read believed, with Wordsworth, that "the child is father to the man" - in other words that our childhood experiences deeply shape and influence the adults we become. He also believed passionately in educating emotions and feelings as much as the intellect. He would have abhorred the extent to which teaching in England has marginalised the arts in favour of pressurised exam-dominated curricula. In his understanding of the role of imagination and creativity in learning and motivation, and the need for different forms of democratic political agendas, much of Read's writing is as acutely relevant for our time as when it was written.

Despite his long residence elsewhere, the North Yorkshire Moors remained a powerful influence in his life and work. In 1935 he produced his remarkable novella, 'The Green Child', written in a genre we now call magic realism. His hero, Olivero, after becoming President of an obscure South American mountain republic through a series of accidents, returns to his native Yorkshire, a thinly disguised Ryedale. He follows an amazing reverse-flowing stream to an old watermill - based on the mill owned by Read's uncle at Hold Caldron in Kirkdale - and rescues a mysterious Green Child. Together they follow the stream to a deep pool on the moors to enter a mystical crystal cavern, where they find a republic based on Read's own anarchistic ideals.

Herbert Read also came home for good in 1949, settling in Stonegrave, in the Howardian Hills, to continue his writing and teaching. He spent the rest of his life there. One of his favourite walks was along Hodge Beck to Bransdale Mill (now a National Trust-owned bunkhouse) which he described as his "spiritual hermitage, the 'bright jewel' to which I often retire in moods of despair". He died on 12th June 1968 and is buried with other members of his family in the graveyard of St Gregory's Minster, in Kirkdale, with its famous Anglo-Saxon sundial and inscription, so beautifully recalled in his poem 'Kirkdale':

I, Orm the son of Gamal
found these fractur'd stones
starting out of the fragrant thicket.
The river bed was dry.

Find out more

St Gregory's Minster, Kirkdale, Kirkbymoorside

There is no better introduction to Herbert Read's Yorkshire than the fine anthology of his writing 'Between the Riccall and the Rye', (Orage Press 2011).

'The Innocent Eye' by Herbert Read, is an autobiography of his childhood in Ryedale (first published 1933) which forms part of the "Annals of Innocence and Experience".

St Gregory's Minster, Kirkdale

Sir Herbert Read

Saxon sundial, St Gregory's Minster

Gravestone, St Gregory's Minster

Boulby Potash Mine, near Staithes

1973

The Story of Boulby Potash Mine

Peter Woods

The mid-19th century saw a boost to the Industrial Revolution in northeast Yorkshire, when Messrs. Bolckow and Vaughan (B&V) stumbled upon near-horizontal beds of iron ore in the Eston Hills, to the south of the River Tees, in 1850.

Events accelerated fast, commencing with the establishment by B&V of an iron-works in what came to be the town of Middlesbrough. The iron-steel works needed fresh water, as opposed to the tidal (and therefore saline) water of the Tees estuary. The new company directors decided that a borehole on their own ground would be the safest bet, so commenced drilling in 1867. The drilling was relatively straightforward for the first 750 feet, but with little sign of water. Then, to everyone's initial pleasure, copious quantities of water were intersected at less than 1,000ft depth. But soon pleasure turned to despair; the water was in fact saturated saline brine, in a 100ft-thick bed of pure salt.

The next chapter in the story was the realisation by local entrepreneurs that this salt-bed had huge industrial potential. Several new boreholes were drilled to exploit the salt layer by 'solution mining' (when water is pumped underground to dissolve the salt and then returned to the surface). Salt extraction led to patchy surface subsidence in the area now known as Saltholme. Initial dismay at this turned to environmental approval: the whole area, with its many shallow freshwater ponds, has become an important bird reserve.

The salt (halite) and its associated beds ('evaporites') were formed by the evaporation of a vast sea which stretched from Yorkshire to Poland and beyond over 250 million years ago. These were the founding resource for ICI and other chemical companies which became large enterprises over the next century.

The next significant chapter began in 1938, when a borehole was drilled for oil and gas at Sleights, near Whitby. It was remarkable because it intersected, at much greater depth, the same 'evaporite' beds, which here included potassium salt layers, which did not occur at Saltholme. ICI had already become a major producer of agricultural fertilisers and recognised the potential of these beds - if only the expertise could be found to reach the depths at which they occurred (over 3,000ft).

World War 2 now intervened, and the new potassium minerals - including the newly discovered polyhalite - were treated for the time being as geological curiosities. After the war, interest in extracting them resumed. Salt-beds traditionally are regarded as potential traps for oil and gas reservoirs beneath them, and agreements were therefore reached in the late 1950s and early 1960s to allow oil companies to deepen any boreholes drilled by ICI in an area from Newton Mulgrave to Loftus, where the potash (i.e. potassium-bearing) beds were less deep than those at Sleights.

The geological guesswork turned out to be sufficiently accurate to allow ICI to search for a partner familiar with sinking deep shafts. The partner chosen, in 1967, was Charter Consolidated Limited (CCL), which took over the exploration drilling programme and established an office at Easington, near Loftus.

Several more boreholes were drilled, which established that the potash beds were continuous, extending out under the North Sea. These showed that the likely reserves would be sufficient to justify a major mine. Considerable thought was given to the best place for siting it, with its two deep shafts (over 3,500ft), its large processing plant, access to the sea for salt-water for the processing, and access to transport for over a million tonnes of potash product per year to a deep-water port. The site chosen was close to the sea at the old Alum mining hamlet of Boulby, west of Staithes, which was also on the recently closed coastal railway. Securing planning permission, in an area of high unemployment following the closure of the iron ore mines, was relatively straightforward. Work began on sinking the shafts in 1969, almost exactly a century after the discovery of the salt-beds on Teesside, and commercial potash production began in 1973.

Ownership was transferred to Israeli Chemicals Ltd (ICL UK) in 2002. The mine now covers 200 hectares and produces half the UK's output of potash, used as an agricultural fertiliser. The rock-salt extracted as a by-product is used for gritting roads in winter conditions.

The site also hosts the Underground Science Laboratory, at 1,100m deep, where experiments can be carried out free from the cosmic radiation which constantly bombards the surface of our planet. The Laboratory is run by the Science & Technology Facilities Council in partnership with ICL UK.

Find out more

ICL UK, http://www.icl-uk.uk/

Boulby Underground Laboratory,
https://www.stfc.ac.uk/about-us/where-we-work/boulby-underground-laboratory/

1973

Shandy Hall and Laurence Sterne

Chris Pearson & Patrick Wildgust

Shandy Hall was built around 1430 as a medieval long hall in the village that was then called Cuckwold. Originally named High Hall, it is sited on high ground at the west end of the village and was originally a timber-framed house with an open hall, central hearth, and solar (an upper-storey room often found in medieval manor houses and used as the family's private living area and sleeping quarters).

Many original features exist, including medieval wall-paintings, and the Grade 1 listed building has had architectural additions in every century; every side of the building looks different. The first family to live there were the Dayvilles, who were then as significant in the area as the Bellasis family of Newburgh Priory. Later families include those of James Hartas (town bailiff); George Spensley (surgeon) and the Reverends Thomas Newton (father and son). In the 19th century the house was divided into two dwellings, and just before it was turned into a museum the Dales and the Smedleys lived there.

The most celebrated inhabitant was Laurence Sterne, who came to live in Shandy Hall in 1760. Sterne was born in Clonmel, Ireland, in 1713. His father was an ensign in the army and in his first years Laurence moved from barracks to barracks. At ten years old he went to school in Hipperholme, Halifax, under the patronage of an uncle, and after his uncle and his father died he was supported by cousins to study at Jesus College, Cambridge, winning a scholarship founded by his great-grandfather, Richard Sterne, a former Archbishop of York. He graduated in 1737 and was ordained into the Church of England as a deacon in the same year. With the help of another uncle, Dr. Jaques Sterne, Precentor of York, he began to make a moderately successful ecclesiastical career. He was ordained priest in 1738 and granted the living of Sutton-on-the-Forest, to which he added the living of Stillington six years later. He married Elizabeth Lumley in 1741, but their daughter Lydia was the only one of several children to survive infancy.

Sterne had already published the first two volumes of 'The Life and Opinions of Tristram Shandy, Gentleman' when he came to Coxwold and his ambition for success ("I write not to be fed, but to be famous") was immediately realised. He became famous virtually overnight and, with a portrait painted of him by Joshua Reynolds within the first few months of his book's release, a celebrity.

His friends celebrated his success by christening his new home 'Shandy Hall', the word 'Shandy' being a dialect word for 'wild', 'merry', and 'odd'. He made frequent visits to London but found the peace he needed for writing at Shandy Hall, living there until his death, and writing the subsequent seven volumes of 'Tristram Shandy' and 'A Sentimental Journey through France and Italy' in a room which now contains the principal part of the museum's collection.

Sterne had been afflicted with illness – a form of tuberculosis - throughout his life. He travelled to France in 1762 to try to improve his health and although he returned to England, his wife and daughter remained in Montpellier. In last years of his life he fell in love with Eliza Draper, a married woman half his age, and wrote 'A Journal to Eliza' when she returned to India with her husband in 1767. He made a "sweet little apartment" at Shandy Hall for her, in the hope that she would one day come there as his wife and muse. She never did. Sterne died in 1768, and was buried three times: once at St George's, Hanover Square, London; secondly when he was recognised after being disinterred by grave robbers and used in an anatomy lecture in Cambridge; and thirdly, when development took place in the 1960s on the London burial ground, his skull and a femur were taken to Coxwold and buried outside the church where he used to preach.

The living of Coxwold had been presented to Sterne by Lord Fauconberg of Newburgh Priory, and Shandy Hall remained in the Bellasis (later Wombwell) family until 1968 when ownership was transferred to the Laurence Sterne Trust. The house was restored and was opened as a museum by comedian and writer Frank Muir in 1973. Along with its two acres of gardens it is now open to the public, and holds exhibitions and events to encourage knowledge and appreciation of Sterne's life and work.

Find out more

The Laurence Sterne Trust,
www.laurencesternetrust.org.uk.

Books by and about Laurence Sterne can be purchased through the website.

Shandy Hall, Coxwold

Revd. Laurence Sterne

Laurence Sterne's study

Shandy Hall, Coxwold

The gardens, Shandy Hall

The Millennium Stone, Danby High Moor

2000

The Millennium Stone on Danby High Moor

Peter Woods

Around the middle of 1999, a member of NYMA asked: "What are we planning to do to mark the Millennium? Shouldn't we build some sort of monument at the highest point in the centre of the Moors?"

The NYMA Council's reaction was that it would never get planning permission for such a structure, but after further deliberation concluded that if the example of Bronze Age standing stones were to be emulated, then the chances of obtaining permission would be higher. And so it turned out; both the Planning Authority and the landowners (Lord and Lady Downe) gave enthusiastic support for a stone to be erected at a spot just off the road between Ralph Cross and Rosedale Abbey.

The next step was to find a suitable stone. Fortunately, the melting ice sheet, some 10,000 years ago, had deposited its load of suitable sandstone close to a road over Spaunton Moor. Another landowner (George Winn-Darley) was enthusiastic too, and so the project had begun.

A local crane company agreed to join the project and move the stone for free, and on 16th February 2000, on a bright but bitterly cold day, the stone - weighing over seven tons - was lifted onto a strong, tractor-drawn trailer and transported ten miles to the site. A large hole had been excavated and lined with stone slabs for long-term stability When the cavalcade arrived, the crane lifted the stone off the trailer and lowered it gently into the hole; it fitted perfectly.

The team then retired to the Lion Inn on Blakey Ridge for a celebration in a nice warm room. Then the snow arrived, turning the new 'standing stone' white. A local farmer, bringing his wife back from a shopping expedition, drove past the stone and said to her, "That's new, isn't it?" "No," she replied, "that's been there as long as we've lived here". When they got home, he turned on the TV and there was the team, raising the new stone. "I got me apology," the farmer reported later! It showed that the decision to emulate the Bronze Age model had been the right one.

It was agreed that the stone should be simply engraved with the letters 'AD MM' divided by a simple quadrant. But when the local stonemason, Mike Weatherill, was asked if he would do the work, he initially declined. A good Methodist, he said: "If I am going to put a cross on the stone, it must be a proper one - are we not celebrating 2000 years since the birth of Jesus Christ?" So that is what is carved on the stone, in perpetuity.

The next step was a dedication ceremony. The much loved and respected local Catholic Priest, Father

Peter Ryan, a true countryman, agreed to lead the ceremony, on 6th May 2000. People wanting to attend on foot were asked to walk from all the points of the compass; many did so (including Lord and Lady Downe). The longest distance walked (15 miles) was achieved by two individuals.

Father Ryan's dedication address included the following apposite words:

"Even I find myself smiling inwardly at the sound of an Irish voice here on the North York Moors; but all the more privileged, very honoured, to be asked to dedicate this beautiful Millennium Stone. It has borne silent, passive witness to many, many millennia, and for a long time it has lain on Spaunton Moor, unnoticed, unheralded. Now it is to bear a different kind of witness. We hope that this Millennium Megalith will be a 'Sermon in Stone' for many generations to come. I hope that it will speak to them, especially through the sign of the cross, about those things that we greatly prize. The cross is a familiar symbol to those of us who are Christians, but in a sense it has real meaning for everyone, irrespective of religious beliefs. In every life, and in creation itself, and certainly in the natural world, there are many dyings. And our life and its worth are measured by the number of 'resurrections' that we can bring about from the many dyings. We hope that this Stone will speak to people about newness of life, about resurrection of life, for many, many generations to come."

Not long after the stone was erected, a palaeontologist contacted NYMA to report that the mud-filled depressions on the back of the stone were in fact small dinosaur footprints in the original sandy beach from which this hard sandstone layer had evolved over millions of years.

Find out more

Visit the Stone on Danby High Moor, just off the road between
Blakey Ridge and Rosedale Abbey, grid ref. NZ 693013

Steve Estill's 2017 book, 'Stones and Crosses of the North York Moors',
published by Fonthill Media, is an excellent guide to the Millennium Stone and other markers.

2006

A Royal Anglo-Saxon Cemetery at Loftus

Steve Sherlock

The Anglo-Saxon period used to be known as the 'Dark Ages', a period after the Romans left Britain, now considered to begin between AD 410 and AD 420. The native people who lived in Britain, who could consider the Celtic tribes as their forebears, continued to tend crops and rear livestock across northeast England. At this time, from the 5th century, a period of Anglo-Saxon migration occurred with people from Angeln in North Germany and southern Denmark settling on the East Coast between Northumberland and East Anglia. These first generations of migrants to our shores were pagan in faith, and cemeteries associated with this period are known at Norton, Saltburn and Hornsea.

In the 7th century the small groups of settlers and native Britons had formed tribal alliances, and Christianity began to emerge at the same time as kingdoms were developing. In this period of change new beliefs emerged and changing practices of burial can be seen with the emergence of 'Conversion Period' cemeteries.

The most significant Conversion Period cemetery in northeast England was excavated at Street House, near Loftus, between 2005 and 2007. The ritual of burial associated with objects changes at this time. In the 6th century weapon burials are associated with some male graves, while the type of objects found in female graves include bronze annular brooches and sleeve clasps, and large numbers (over 100) of amber beads. By the 7th century graves contained fewer objects, and glass beads are found in small clusters along with ornate pendants, worn on a chain, such as the Street House example.

In the centre of the Street House cemetery was a burial that contained a significant amount of ironwork around the edge of a burial chamber, with three gold pendants. This burial is the only known Anglo-Saxon 'bed burial' in northeast England. The term relates to the fact that an individual is interred lying upon a bed. This type of burial dates from AD 640–660 and is a rite known after the conversion from pagan practices to Christianity. In total there are 17 burials of this type currently known in England, mostly in East Anglia and southern England from Dorset and Wiltshire, and they are a type of female burial, with one exception, which may be an anomaly. It is suggested by some people that high-status royal females are placed on a decorated bed whilst males of similar status are buried with weapons, drinking vessels, horses, or perhaps a boat as indicators of esteem.

Within the group of known bed burials the contents of the grave can vary. Whilst some contain few finds, the Street House burial contained a number of high-status objects, but it also differed in several respects. The bed was made of ash and the ironwork, although similar in function to that of other examples - with headboard stays, plates and nails - was much more ornate. The grave also contained two gold cabochon pendants (just one would be rare in any grave) and the Street House gold pendant.

The Street House pendant is unique in the Anglo-Saxon world. It is made of gold and measures 37mm by 27mm. The pendant has a loop for suspension on a chain and it is a piece of exceptional craftsmanship. There are 57 small cells containing reused garnets cut and shaped with a degree of skill comparable with jewellery from Sutton Hoo in Suffolk. The centrepiece of the pendant is a larger gemstone with a number of incised lines that form the shape of a scallop shell. The scallop is a motif occasionally found on Christian sarcophagi and is associated with the Apostle St James. Clearly, whoever was the owner and wearer of the pendant had wealth, status and access to the best craftsmen in Anglo-Saxon England. The jewellery, allied to the burial chamber, bed and uniqueness of the find in northeast England, indicates that the person buried at Street House was of high status: hence the term 'Anglo-Saxon Princess'.

The discovery of the Street House Anglo-Saxon cemetery has led to many further questions about who the Princess was, where she lived and what she was doing at Street House. Archaeological research at Street House continues in the search for an Anglo-Saxon village in the immediate area of the cemetery.

Find out more

The finds from the excavation can be seen at Kirkleatham Museum, Kirkleatham, Redcar TS10 5NW, http://www.redcar-cleveland.gov.uk/kirkleathammuseum

Saxon Princess Trail (Loftus): Self-guided walk leaflet available at https://www.walkingloftusandthenorthyorkshirecoast.com/self-guided-walks

The Street House pendant (gold and garnets)

Gold filigreed pendant found at Street House

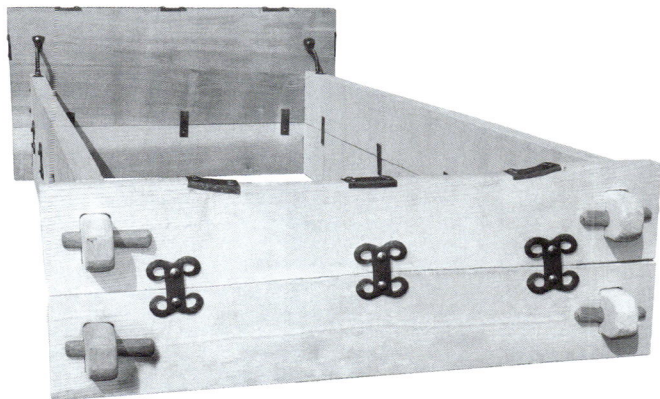

Replica of ash-wood bedstead used in Loftus bed-burial

THE HISTORY TREE

This stump is all that remains of a once majestic copper beech tree. The iconic tree was planted circa 1800 and flourished here for over 200 years, living through the reigns of nine British monarchs. For over two centuries the tree was mute witness to great change and many events occurred during this period that are an important legacy of the rich culture and social heritage of the North York Moors and adjacent areas. A varied selection of these historical events has been chosen to feature on this plate.

Positions of the event dates on the plate have been calculated from the annual growth rings showing on the face of the ancient tree stump.

Trees are living landmarks, a link with the past and a symbol of hope for the future. They grow larger and live longer than anything else on earth. They adorn our landscapes, contribute to a healthy and sustainable environment, provide a haven for wildlife, and are a valuable natural resource. Trees enrich our lives, bring us closer to nature, and are vital to the future survival of mankind.

"If a tree dies, plant another in its place."
Linnaeus 1707 - 1778

Inscription on the The History Tree Plate at the Moors Centre

About The Authors

Beth Andrews is a geologist with an interest in industrial archaeology. She worked as Geodiversity and Heritage Officer with the Tees Valley Wildlife Trust and was Chair of the Tees Valley RIGS (Regionally Important Geological and Geomorphological Sites) Group for 8 years.

Sharon Artley taught at Caedmon School in Whitby for 23 years. When a replica of Caedmon's Cross was erected in the school grounds, she researched how the original Cross came into being and learned Caedmon's hymn in Anglo-Saxon. Since 2018 Sharon has been the principal editor of NYMA's quarterly magazine 'Voice of the Moors'.

Fiona Barnard is a life-long history enthusiast whose interest in the Scoresbys was sparked when she moved to Whitby. She has been working on the Scoresby collection in Whitby Museum since 2009.

Ian Carstairs has wide experience of conservation of the natural, cultural and built heritage, charity trusteeships, community action and campaigning. Starting his countryside 'career' as Assistant Director of the Moors Centre, he has also been a Secretary of State's Board member and a deputy-chair of the North York Moors National Park Authority; Chairman of the Heritage Lottery Fund's Yorkshire and the Humber grant committee; and is President of NYMA. A competent photographer, he loves the moorland landscape and the night sky.

Michael Chaloner is a retired Industrial Chemist and treasurer of Northallerton and District Local History Society. These are linked by an interest in the alum industry of North Yorkshire, the beginning of Britain's chemical industry, and links to his family, which developed the first alum mines in England.

Linda Chambers was instrumental in starting the Rosedale History Society in 2008 and since then has been working to collect and make available the growing archive of the area's varied local and industrial history. Involvement in several history-based projects including the 'Lord of Iron' programme has enabled her to gain invaluable insight into the diversity and historical importance of the wider area.

Janet Cochrane has had a life-long interest in wild places, especially in experiencing them from the back of a horse. Her career has principally been spent in the travel industry and academia, with a particular interest in resilient tourism systems in national parks and other rural areas, and including 12 years as Senior Research Fellow in Responsible Tourism Management at Leeds Beckett University. She became Secretary of NYMA in 2015.

Albert Elliot has been an advocate of the North Yorkshire Moors ever since visiting the beautiful moorland, dale, and coastal areas on a solo cycling holiday as a teenager in 1956. He returned time after time, cycling, hiking and rock climbing, often staying at youth hostels where he made many lifelong friends. He developed an abiding interest in the area's history, social and cultural folklore, and now resides within the National Park. He was editor of NYMA's quarterly magazine 'Voice of the Moors' for 9 years until 2018.

Jane Ellis has a lifelong passion for the North York Moors and has been researching and photographing their curiosities for many years, with particular reference to the industrial history of the region. She is a volunteer at Robin Hood's Bay Museum and an active member of several Yorkshire cultural and natural history societies and walking groups.

Sheila Gendle-Clarke and Ray Clarke have spent a number of years researching their family trees. When they moved to Scarborough in 2009 they discovered that Sheila's late brother-in-law, Frank Graham, was a great-great-great-nephew of Henry Freeman.

Ann Glass was born in East Yorkshire and spent childhood weekends and holidays in the North York Moors. Since the 1990s she has spent increasing amounts of time there with her family. She now enjoys the Moors every day, living in and conducting her legal practice from Rosedale.

Adrian Leaman has a lifelong interest in landscape and vernacular buildings. His professional life has been concerned with studying how people use buildings and with helping designers and managers improve building performance for the occupants.

Rita Leaman is a psychotherapist and writer who lives in North Yorkshire. As Alison R. Russell, she is the author of 'Are You Chasing Rainbows?' (www.chasingrainbows.org.uk) and writes a blog on emotional health. Her work focusses particularly on emotional needs that are unmet and needs met unhealthily; this is why she responded favourably to the Camphill Trust, where the clients' needs are met well.

Shirley Learoyd was born within sight of the White Horse, in the School House in Coxwold, and has had a continuing love of and interest in the area ever since.

Carolyn Moore has enjoyed years of visiting and now living in the North York Moors. She has a particular interest in Captain James Cook, having grown up in his birthplace village and visiting his monument on Easby Moor regularly for the past 50 years.

Sue Morton is an architectural and landscape painter based near Whitby and a member of the Whitby Art Society. In 2013 RNLI Cromer asked the Society to create a mosaic of 64 paintings depicting the 1914 wreck and rescue of HMHS Rohilla. This became part of an acclaimed national travelling exhibition (2014-2019) entitled 'Hope in the Great War'.

Louise Mudd is a busy working mother of two, whose involvement in local history began when she took over the role of secretary of the Kirkbymoorside History Group. Her particular interest is in researching and bringing back to life the forgotten stories linked to local places and families.

Tamsyn Naylor 's interest in the history of the Esk Valley was sparked by members of the Cleveland Mining Heritage Society. Living surrounded by such a rich heritage and working for the steam railway at Grosmont meant she was ideally placed to get involved with the National Park's 'This Exploited Land of Iron' project, as well as contributing articles on a variety of topics in local magazines such as 'Esk Valley News'.

Adele Pennington was born with a spirit of adventure. Climbing Mount Snowdon at the age of 8 sparked a dream to reach the world's highest mountain. While studying for a PhD at York University she became a member of the Scarborough and District Mountain Rescue Team and spent all her free time climbing. She eventually gave up teaching to concentrate on outdoor pursuits, and now runs her own mountaineering company.

Andrew Scott has been a North Yorkshire Moors Railway volunteer since its inception nearly 50 years ago. Having retired from his role as Director of the National Railway Museum in 2010, he now devotes much of his time to the NYMR and to the National Park, of which he is a Board Member.

Mike Shaw is curator of the Sutcliffe collection of photographs and owner of the Sutcliffe Gallery in Whitby, where many of them are displayed. He has published several compilations of Frank Meadow Sutcliffe's photos.

Steve Sherlock has been a professional archaeologist working in Northeast Yorkshire since the 1970s. His research interests are wide although he has specialised in the Iron Age through to the Anglo-Saxon period. His present research is focused on the landscape around Loftus, on the Yorkshire Coast.

Colin Speakman is best known for his writing about the Yorkshire Dales, but is also a recognised poet and authority on rural transport and sustainable tourism. He has had a long-standing interest in Wordsworth, the first great English Romantic walker-poet, and Herbert Read as a poet, art critic and philosopher.

Fleur Speakman's interest in the Gothic and horror novel genre was kindled originally by a university course, emphasising psychological and sociological insights. An abiding interest in landscape and the historic past have featured prominently in a career of teaching, editorial work and consultancy in sustainable and heritage tourism.

Alan Staniforth was an Information Officer and Heritage Coast Ranger in the North York Moors National Park for 30 years. William Smith has been a life-long hero of his, not least because of Smith's humble background and his long fight for recognition.

Brian Walker was raised in Scarborough and joined the Forestry Commission in 1976, working in the North Pennines until returning to North Yorkshire in 1994 to manage recreation, access and the environment in the Commission's Yorkshire forests. Since retiring in 2010 Brian has been active in Scarborough Field Naturalists and is a member of the National Park's team of volunteers.

Patrick Wildgust took over in 2002 as curator of Shandy Hall and he and his partner Chris Pearson are lucky enough to live there. Between them they have built up the educational side of the Laurence Sterne Trust and manage both the house and the two-acre garden.

Carol Wilson has a keen interest in local history. After retiring as a head-teacher she went on to take a Masters' degree in medieval studies at the University of York, to research the village of Westerdale, and to write the history of Castleton School.

Mike Windle has a lifelong fascination for our planet. As a geologist he has spent the last 30 years sharing his passion through practical conservation, as director of two geology trusts and as an educator with his Geonaut programme. Lewis Hunton is one of his scientific heroes.

Elaine Wisdom lives on the edge of the North Yorkshire Moors and the coast. Both are ancient habitats for human life, from the neolithic through industrialisation to the present day. She enjoys engaging with that sense of lived continuity within the rugged panorama of both moors and sea, through her voluntary work with Ryedale Folk Museum and with the NYMA walking group.

Peter Woods studied Geology at Cambridge University in the 1950s, and after working in Africa moved to Yorkshire to help develop the new potash mine at Boulby. At the age of 50 he did an MSc in Forestry and Wildlife Habitat Management and then worked as an environmental management consultant. He was involved in raising the Millennium Stone while Secretary of NYMA in 2000.

Useful Addresses

North Yorkshire Moors Association (NYMA)
4 Station Road
Castleton
Whitby
North Yorkshire YO21 2EG
https://www.north-yorkshire-moors.org.uk/
(Please note that NYMA does not have
premises open to the public)

North York Moors National Park Authority
The Old Vicarage
Bondgate
Helmsley
North Yorkshire YO62 5BP
http://www.northyorkmoors.org.uk/

Sutton Bank National Park Centre
Sutton Bank
Thirsk
North Yorkshire YO7 2EH

The Moors National Park Centre
Lodge Lane
Danby
Whitby
North Yorkshire YO21 2NB

Campaign for National Parks
Europoint House
5 Lavington Street
London SE1 ONZ
https://www.cnp.org.uk

About The North Yorkshire Moors Association

NYMA is dedicated to safeguarding the landscape and culture of the North York Moors National Park and surrounding areas. It supports development which respects the special character of the Moors while taking account of the needs of residents, working communities and visitors.

If you care about and enjoy this spectacular moorland landscape and coast, why not become a member?

Find out more about our work, campaigns, publications, walks and events at:

https://www.north-yorkshire-moors.org.uk/

Information and free downloads about The History Tree project,
including the learning pack, can also be found on our website.

Acknowledgements

The contributions of at least 40 people have made this publication possible, not least the 30 authors who have researched and written the topics which form the main body of the book.

Amongst these, principal thanks must go to Albert Elliot who, as a stalwart of the North Yorkshire Moors Association over many years and a long-term resident of the National Park, identified the opportunity to commemorate the copper beech tree at Danby Lodge after it was felled in 2007. It was he who master-minded the design and installation of the History Tree plate which serves both as a point of interest at the National Park's Moors Centre and as a recurring motif in our book.

We are also hugely grateful for the creative insights and many hours of work that NYMA President Ian Carstairs has put into supporting the production of this volume, helping with the photographs, urging the team forward when spirits were flagging, and drawing together the final strands of the various elements of the book.

It is certain that without the quiet dedication to high standards and speedy working of our designer, Ian Dashper, the book would not have reached the quality that it has.

Other essential support for the project has come from Carolyn Moore, who led on producing the educational materials which complement the book, and Brian Pearce, NYMA's Treasurer, who has helped to ensure that the project achieved its aims within the grants awarded by the Heritage Lottery Fund and the Land of Iron project.

Special thanks are also due to Whitby Museum, The Rotunda Museum, Ryedale Folk Museum, and Robert Thompson's, Kilburn whose staff cheerfully accommodated at short notice requests to photograph exhibits in their collections.

Photographs and Illustrations

Credits are listed in alphabetical order of ownership. On pages with multiple images the acknowledgements refer to them by Roman numeral, clockwise from top left (reference numbers are not shown with the individual images). The contributors directly involved all gifted use of their material free of charge to NYMA for this publication.

By courtesy of Emma Beeforth: 97(i-ii); Phil Brown: title page; William Wordsworth by William Shuter ©The Division of Rare and Manuscript Collections, Cornell University: 14(ii); Debra Decloux: 49(ii); Dorman Museum: 67(i); Tim Dunn: 49(iii); Albert Elliot: 7,19(iv),back cover; William Smith by Hugues Fourau ©The Geological Society: 32(i); R.J. Hammerton after a photograph by Whitby Wallis; North York Moors National Park Authority: 13; 5; Raymond Hayes Collection/Ryedale Folk Museum: 97(iii); Mike Kipling: ©front cover; Neil Mitchell – APS(UK)/Carstairs Countryside Trust: 85; Captain James Cook by Nathaniel Dance ©National Maritime Museum: 25(ii); Sir Herbert Read ©National Portrait Gallery: 121(ii); Sir George Cayley by Henry Perront Briggs ©National Portrait Gallery: 44(ii); Adele Pennington: 104(i); Alan Staniforth: 14(i),(v),26(iii),37(i)(iv),43(i),49(iv),62(i),68(ii),73(iii),74(ii),80(ii),91(i-iii),110; Steve Sherlock/Kirkleatham Museum: 133; The Sutcliffe Gallery: 79,92(i-iv); Robert Thompson's Craftsmen Ltd.: 62(ii); Whitby Museum: 26(ii); by courtesy of Vicky Whitfield: 97(iv-v). All other modern photographs: Ian Carstairs. We were unable to find sources for certain historic images.